Advances in Numerical Analysis Emphasizing Interval Data

Advances in Numerical Analysis Emphasizing Interval Data

Tofigh Allahviranloo
Witold Pedrycz
Armin Esfandiari

CRC Press
Taylor & Francis Group
Boca Raton London New York

CRC Press is an imprint of the
Taylor & Francis Group, an **informa** business

First edition published 2022
by CRC Press
6000 Broken Sound Parkway NW, Suite 300, Boca Raton, FL 33487-2742

and by CRC Press
4 Park Square, Milton Park, Abingdon, Oxon, OX14 4RN

Library of Congress Cataloging-in-Publication Data
A catalog record for this title has been requested

ISBN: 978-1-032-11043-1 (hbk)
ISBN: 978-1-032-11045-5 (pbk)
ISBN: 978-1-003-21817-3 (ebk)

DOI: 10.1201/9781003218173

Typeset in Palatino
by codeMantra

Contents

Preface

Numerical analysis is the cornerstone of computations in numerical methods, especially recently, interval numerical calculations play an important role in all topics of engineering and physical sciences and even in life sciences, social sciences, medicine, business, and art. The recent growth in computational potency has made it possible to use more sophisticated numerical analysis and to provide accurate and realistic mathematical models in the sciences. Researchers' interest in computing uncertain data, that is, interval data, opens up new avenues for tackling real-world problems and offers innovative and efficient solutions.

This book provides the basic theoretical foundations of numerical methods, discusses classes of key techniques, explains their advances, and provides insights into recent developments and challenges. The theoretical parts of numerical methods, including the concept of interval approximation theory, are introduced and explained in detail. In general, the first feature of the book is an up-to-date and focused dissertation on computational error analysis, in particular a comprehensive and systematic treatment of error propagation mechanisms. The second one is considerations on the quality of data related to numerical calculations with a full discussion of distance approximation theory. The third feature focuses on the theory of approximation and its development from the perspective of linear algebra. Finally, the new and regular view of numerical integration and their solutions enhanced by error analysis is presented as the last feature of the book.

The book will be of interest to the broad spectrum of readers exploring interval computing in numerical analysis. It could be also of particular interest to graduate students in engineering and applied mathematics.

Authors

Tofigh Allahviranloo is a full professor of applied mathematics at Istinye University, Turkey, with more than 250 publications on topics of numerical analysis, 11 books in Farsi, three books published by Elsevier and Springer, and 20 contributions to conference proceedings, accumulating close to 8200 Google Scholar citations (h-index 46) uncertainties and soft computing. Prof. Allahviranloo is the associate editor in charge for the *Journal of Information Science, Fuzzy Sets and Systems*, the *Journal of Intelligent and Fuzzy Systems*, the *Iranian Journal of Fuzzy Systems*, and the *Journal of Mathematical Sciences*. He is also editor-in-chief of the *International Journal of Industrial Mathematics*. Tofigh has developed a passion for multi- and interdisciplinary research. He not only pursues fundamental research in fuzzy applied mathematics, especially fuzzy differential equations, but also develops innovative applications in the applied biological sciences.

Witold Pedrycz (IEEE Fellow, 1998) is professor and Canada Research Chair (CRC) in Computational Intelligence in the Department of Electrical and Computer Engineering, University of Alberta, Edmonton, Canada. He is also a member of the Systems Research Institute of the Polish Academy of Sciences, Warsaw, Poland. In 2009, Dr. Pedrycz was elected as a foreign member of the Polish Academy of Sciences. In 2012, he was elected as a fellow of the Royal Society of Canada. In 2007, he received a prestigious Norbert Wiener award from the IEEE Systems, Man, and Cybernetics Society. He is a recipient of the IEEE Canada Computer Engineering Medal, a Cajastur Prize for Soft Computing from the European Centre for Soft Computing, a Killam Prize, a Fuzzy Pioneer Award from the IEEE Computational Intelligence Society, and 2019 Meritorious Service Award from the IEEE Systems Man and Cybernetics Society.

His main research directions involve computational intelligence, fuzzy modeling and granular computing, knowledge discovery and data science, pattern recognition, data science, knowledge-based neural networks, and control engineering. He has published papers in these areas. He is also an author of 21 research monographs and edited volumes covering various aspects of computational intelligence, data mining, and software engineering.

Dr. Pedrycz is vigorously involved in editorial activities. He is an editor-in-chief of *Information Sciences,* editor-in-chief of *WIREs Data Mining and Knowledge Discovery* (Wiley), and co-editor-in-chief of *Int. J. of Granular Computing* (Springer) and *J. of Data Information and Management* (Springer). He serves on an Advisory Board of *IEEE Transactions on Fuzzy Systems* and is a member of a number of editorial boards of international journals.

Armin Esfandiari is a PhD candidate and researcher at Istinye University.

1

About the Book

Numerical analysis forms a cornerstone of numeric computing and optimization, in particular recently, interval numerical computations play an important role in these topics. The interest of researchers in computations involving uncertain data, namely, interval data, opens new avenues in coping with real-world problems and delivers innovative and efficient solutions.

The book delivers essential theoretical fundamentals of numerical methods, discusses classes of key techniques, elaborates on their refinements and enhancements, and offers insights into the recent developments and existing challenges.

The theoretical parts of numerical methods including the concept of interval approximation theory are introduced and explained in detail. In general, the key features of the book include:

- Updated and focused treatise on error analysis in computing, especially a comprehensive and systematic treatment of mechanisms of error propagation.
- Considerations on quality of data involved in numeric computing with a thorough discussion of interval approximation theory.
- Focus on approximation theory and its development from the linear algebra point of view.
- A new and well-organized exposure of numerical integration and their solutions enhanced with error analysis.

2

Error Analysis

DOI: 10.1201/9781003218173-2

2.1 Introduction

Based on the discussed numerical methods and considering other chapters of the book, we need to explain the error analysis first. In addition to exploring the numerical integration methods, the interpolation of integrand function is needed and this is one of the reasons to discuss about interpolation methods.

2.2 Error Analysis

In this chapter, we intend to investigate and analyze the important complicated problems and points that occur in numerical calculations or calculations based on the numerical algorithms. As we know, in the numerical analysis, most numerical methods are iterative. This means that their formulation is in the form of difference equations. Therefore, given one or more initial values, the next values must be calculated. Usually, the initial values are not accurately available and are approximate; or due to the structure of the mathematical model, the calculations performed using iterative methods produce approximate results, that is, whether the initial values are approximate or not, the difference model may also have error factors. Obviously, two points were always considered in the computer or the numerical calculations. One is the speed of calculations and the other is the memory occupied by numerical results. Due to the advances in science and technology, the second factor has been ignored in the presentation of structured algorithms, but the first factor is considered as an advantage for the presentation of numerical algorithms. Given that each computational device has its own computational accuracy, it can be said that the zero of each computing device or computer is different from that of another computer, that is, the smallest positive number of one machine is different from that of another machine. Therefore, an algorithm performed in two machines will have different results. But in both cases, there is a computational error which is less in one than the other. Currently, due to the advances in technology and the construction of advanced satellites and long-range air-to-air missiles and missiles with nuclear warheads, an approximate estimate of the target with the lowest error rate and the calculations of missile or satellite launch with the least amount of error is important. This is because the missile is trying to hit a specified target over a distance of, for example, thousands of kilometers, which may also be approximate. However, how to launch the missile, initial speed, initial acceleration, traveled distance, obstacles in the path of the missile such as air resistance, winds blowing from lateral directions, etc., and how to hit the target, all are factors that required to be considered, and obviously none of these factors can be accessed accurately and without an error. Therefore, taking into account these factors and problems, hitting the target with a missile should be done with an error of, for example, a maximum of 0.01. Obviously, all models related to this process are in the form of mathematical models, for example, the differential equations with the initial conditions, the integral equations, the differential equations with partial derivatives, the calculations of integral series, and so on. So we need to examine the errors of such models and estimate the upper and lower bounds of such errors. In this regard, some problems about error analysis are presented.

2.2.1 Errors in an Algorithm

Suppose that $Y = \phi(X)$, where ϕ is a combination of all the steps of the algorithm. For this purpose, we define:

$$\phi : D \to \mathbb{R}^m$$

where D is an open subset of $\mathbb{R}^n$. We also assume that $X^t = (x_1, \ldots, x_n)$ and $Y^t = (y_1, \ldots, y_m)$ are the input and output vectors of the algorithm, respectively. It is clear that:

$$y_i = \varphi_i(x_1, \ldots, x_n), \quad i = 1, \ldots, m$$

that is:

$$Y = \begin{bmatrix} y_1 \\ \vdots \\ y_m \end{bmatrix} = \begin{bmatrix} \varphi_1(x_1, \ldots, x_n) \\ \vdots \\ \varphi_m(x_1, \ldots, x_n) \end{bmatrix}$$

If we want to specify ϕ for an algorithm that has $r+1$ operators (steps), we have:

$$\varphi^{(i)} : D_i \to D_{i+1}, \quad i = 0, \ldots, r, \ D_i \subseteq \mathbb{R}^{n_i}, \ n_i \in \mathbb{Z}$$

$$\phi = \varphi^{(r)}...\varphi^{(0)},\, D_0 = D,\, D_{r+1} \subseteq \mathbb{R}^{n_r+1} = \mathbb{R}^m$$

To calculate ϕ in each algorithm, we have the ordered series of $\varphi^{(i)}$ operators with the sum equal to ϕ, so that the output of one operator will be the input of the next operator, and finally the output of the last operator will be Y.

If in i th step of the algorithm, the vector $X^{(i)}$ has n_i inputs for the operator $\varphi^{(i)}$, then we have:

$$\varphi^{(i)} : D_i \to \mathbb{R}^{n_i+1},\, D_i \subseteq \mathbb{R}^{n_i}$$

So that

$$\varphi^{(i)}\left(X^{(i)}\right) = X^{(i+1)}$$

We have already mentioned that:

$$\phi : D \to \mathbb{R}^m,\, D \subseteq \mathbb{R}^n$$

$$\phi(X) = \begin{bmatrix} \varphi_1(x_1,...,x_n) \\ \vdots \\ \varphi_m(x_1,...,x_n) \end{bmatrix}$$

We assume that elements of $\phi(X)$, that is, φ_is, have a continuous first derivative and $\tilde{X}$ is an approximation of x. We know that absolute and relative error of $\tilde{X}$ are defined as follows:

$$\Delta\tilde{X} = \tilde{X} - X \tag{2.1}$$

$$\varepsilon_{\tilde{X}} = \frac{\tilde{X} - X}{X} \tag{2.2}$$

Considering the absolute and relative error of $\tilde{X}$, the absolute error and the relative error of $\tilde{Y} = \phi\left(\tilde{X}\right)$ will be as follows:

$$\Delta\tilde{Y} \leq \sum_{j=1}^{n} \left|\frac{\partial\varphi_i(X)}{\partial x_j}\right| \cdot \Delta\tilde{x}_j \tag{2.3}$$

$$= D\phi(X) \cdot \Delta\tilde{X} \tag{2.4}$$

And by using the definition of relative error and formula (2.2)

$$\varepsilon_{\tilde{y}_i} \leq \sum_{j=1}^{n} \left|\frac{\partial\varphi_i(X)}{\partial x_j}\right| \cdot \frac{x_j}{\varphi_i(X)} \cdot \varepsilon_{\tilde{x}_j} \quad x_j \neq 0,\, j=1,...,n,\, y_i \neq 0,\, i=1,...,m \tag{2.5}$$

In relation to (2.4), $D\phi(X)$ is the Jacobian matrix. The factors $\dfrac{\partial\varphi_i(X)}{\partial x_j}$ in (2.3) are called sensitivity numbers and also the factors $\dfrac{\partial\varphi_i(X)}{\partial x_j}\cdot\dfrac{x_j}{\varphi_i(X)}$ in (2.5) are called condition numbers. If in a problem, all condition numbers are small, we say that the problem is well established.

The following problem describes the analysis of the series calculation error with a computational tool. Due to the widespread application of series, it is important to explain such problems.

2.2.1.1 Problem

Calculating $\displaystyle\sum_{i=1}^{n} a_i$ results in an arbitrary large relative error. If all terms of a_i have the same sign, then the relative error is bounded, find a bound for this error, ignoring the higher order terms.

Solution: Suppose that $y = \displaystyle\sum_{i=1}^{n} a_i$ and $\tilde{y}$ is an approximation of it. According to the relative error formula:

$$\varepsilon_{\tilde{y}} \le \sum_{j=1}^{n} \frac{x_j}{\phi(x)}\cdot\left|\frac{\partial\phi(x)}{\partial x_j}\right|\cdot\varepsilon_{x_j}$$

we have:

$$\varepsilon_{\tilde{y}} \le \sum_{j=1}^{n} \frac{a_j}{a_1 + \cdots + a_n}\cdot|1|\cdot\varepsilon_{a_j} = \frac{a_1}{a_1 + \cdots + a_n}\cdot\varepsilon_{a_1} + \frac{a_2}{a_1 + \cdots + a_n}\cdot\varepsilon_{a_2}$$

$$+ \cdots + \frac{a_n}{a_1 + \cdots + a_n}\cdot\varepsilon_{a_n}$$

So

$$\varepsilon_{\tilde{y}} \le \left|\frac{a_1}{a_1 + \cdots + a_n}\right|\cdot|\varepsilon_{a_1}| + \cdots + \left|\frac{a_n}{a_1 + \cdots + a_n}\right|\cdot|\varepsilon_{a_n}|$$

If all terms have same sign, then:

$$\left|\frac{u_i}{a_1 + \cdots + a_n}\right| \le 1, \quad i = 1,\ldots,n$$

And if $\displaystyle\max_{i=1,\ldots,n}|\varepsilon_{a_i}| = M$, we will have:

$$\left|\mathcal{E}_{\bar{y}}\right| \leq \left|\mathcal{E}_{a_1}\right| + \cdots + \left|\mathcal{E}_{a_n}\right| \leq n \cdot M$$

So, the relative error has an upper bound.
 Now if all terms have not same sign, then:

$$\exists i \left(i = 1,\ldots,n \Rightarrow \left|\frac{a_i}{a_1 + \cdots + a_n}\right| \geq 1 \right)$$

and the error will be huge.
 The purpose of expressing and solving the following problem is to point out that the subtraction of close approximate numbers of the same sign has a large relative error.

2.2.1.2 Problem

Consider a quadratic equation. If $b > 0$ and $b^2 \gg ab$, how should the root of the equation be calculated to minimize the calculation error?
Solution: The roots of the equation are:

$$x_1 = \frac{-b + \sqrt{b^2 - ac}}{a}, \, x_2 = \frac{-b - \sqrt{b^2 - ac}}{a}$$

Since $b > 0$ and $b^2 \gg ab$, so the value of $\sqrt{b^2 - ac}$ is very close to b, and there-fore, in the nominator of x_1, we will have a subtraction of two close numbers, so there is a big error for calculating x_1, but in calculating x_2, there is not. So, first we obtain x_2 and then, we also calculate x_1 as follows:

$$x_1 = \frac{-b + \sqrt{b^2 - ac}}{a} \cdot \frac{-b - \sqrt{b^2 - ac}}{-b - \sqrt{b^2 - ac}}$$

$$= \frac{b^2 - \left(b^2 - ac\right)}{a \cdot x_2 \cdot a}$$

$$= \frac{c}{ax_2}$$

2.2.1.3 Definition-Absolute Error

If $\tilde{x}$ is an approximation of x, then

$$e(\tilde{x}) = |x - \tilde{x}|$$

is called an absolute error of a_n.

2.2.1.4 *Example*

1. Suppose that $a_n = \dfrac{n+1}{n}$, what is the absolute error of a_n as an approximation of the number one?

$$e(a_n) = \left|1 - \frac{n+1}{n}\right| = \frac{1}{n}$$

It can be observed that the larger the n, the smaller the $\dfrac{1}{n}$, and as a result, the a_n gets closer to one. If we want the error of a_n to be smaller than, for example, 10^{-3}, it is enough to write:

$$\frac{1}{n} < 10^{-3}$$

Therefore, $n > 10^3$. The first n that holds in the recent inequality is 1001, for which:

$$a_n = \frac{1002}{1001} = 1.000999$$

2. We know that 1.141 is an approximation of $\sqrt{2}$. What is the absolute error of 1.41?

   Obviously, if we write the decimal expansion of $\sqrt{2}$ using a calculator, that is,

$$\sqrt{2} = 1.414213562(9D)$$

then

$$e(1.41) = 0.0042135562000$$

$$2.66 \leq l \leq 2.74$$

Now the question is whether the absolute error of an approximation determines the accuracy of that approximation? Read the following and answer the relevant questions to find out the answer of the above question.

A. Consider two bank cashiers, one of whom has lost one hundred Tomans by exchanging, for example, one million Tomans, and the other has extra one hundred Tomans by exchanging five hundred thousand Tomans.
   Which cashier was more accurate?

B. Consider two goalkeepers, one of whom has conceded four goals from five penalties and the other has conceded six goals from 10 penalties. Which goalkeeper was better?

From the above examples, it follows that what determines the accuracy of an approximation is the error per unit of quantity; the smaller the error, the accurate the approximation. Therefore, we have the following definition.

2.2.1.5 Definition – Relative Error

If $\tilde{x}$ is an approximation of a non-zero number x, the relative error of x is denoted by $\delta(\tilde{x})$, which is the error per unit of the quantity, that is,

$$\delta(\tilde{x}) = \frac{|x - \tilde{x}|}{|x|}$$

In most numerical analysis problems, there is not a real number x. For this purpose, an upper bound can be obtained for $\delta(\tilde{x})$ without the existence of x.

2.2.1.6 Problem

Suppose that $\bar{x}$ is an approximation of x and $\bar{y}$ is an approximation of y. Calculate the relative error $\frac{x}{y}$.

Solution: suppose that $\varepsilon := \bar{x} - x$ and $\eta : \bar{y} - y$. So,

$$\varepsilon_{\frac{x}{y}} = \frac{\frac{\bar{x}}{\bar{y}} - \frac{x}{y}}{\frac{x}{y}}$$

$$= \frac{\frac{x+\varepsilon}{y+\varepsilon} - \frac{x}{y}}{\frac{x}{y}}$$

$$= \frac{\frac{x(1+\varepsilon_{\bar{x}})}{y(1+\varepsilon_{\bar{y}})} - \frac{x}{y}}{\frac{x}{y}}$$

$$= \frac{1+\varepsilon_{\bar{x}}}{1+\varepsilon_{\bar{y}}} - 1$$

$$\approx \varepsilon_{\bar{x}} - \varepsilon_{\bar{y}}$$

2.2.1.7 Theorem

If $\tilde{x}$ is an approximation of x, and $e_{\tilde{x}}$ is an absolute limit error of $\tilde{x}$, we have

$$\delta(\tilde{x}) \leq \frac{e_{\tilde{x}}}{|\tilde{x}| - e_{\tilde{x}}}$$

Proof: According to the hypothesis, we have

$$|x - \tilde{x}| \leq e_{\tilde{x}}$$

and according to the properties of the absolute value,

$$|\tilde{x}| - |x| \leq |x - \tilde{x}|$$

As a result, the sentence is true.

If $e_{\tilde{x}}$ is small compared to $|\tilde{x}|$, it can be ignored and so we can write

$$|\tilde{x}| - e_{\tilde{x}} \approx |\tilde{x}|$$

so, we have the following remark:

2.2.1.8 Remark

If $e_{\tilde{x}}$ is small compared to $|\tilde{x}|$

$$\delta(\tilde{x}) \underset{\sim}{<} \frac{e_{\tilde{x}}}{|\tilde{x}|}$$

Practically, the right side of the above inequality is non-computable, and therefore, assuming that $e_{\tilde{x}}$ is negligible in comparison with $|\tilde{x}|$, the relative error is defined as follows:

$$\delta(\tilde{x}) \approx \frac{e_{\tilde{x}}}{|\tilde{x}|} \tag{2.6}$$

2.2.1.9 Example

Suppose $\tilde{x} = 1.41$ and $x = \sqrt{2}$; The relative error of $\tilde{x}$ is as follows

$$\delta(\tilde{x}) = \frac{\sqrt{2} - 1.41}{\sqrt{2}} = 0.002979438000$$

But if we consider $e_{\tilde{x}} = 0.005$, we will have

$$\delta(\tilde{x}) \approx \frac{e_{\tilde{x}}}{|\tilde{x}|} = 0.003546099291000$$

2.2.1.10 Different Types of Error Sources

According to what is stated, a list of the error sources is presented in the following.

A. **The error of the model.** This error includes omissions, ignorance, and simplifications to determine the mathematical model of the problem.

B. **The error of the data.** This error occurs when measuring and estimating problem assumptions.

C. **The error of the number representation.** Decimal or binary representation of most numbers with finite figures is not possible. Therefore, selecting a finite number of expansion digits of a number causes this error.

D. **The error of the arithmetic operation.** The result of some operations on two factors has an infinite number of digits and selecting a finite number of these digits causes this error.

E. **The error of the method.** Numerical methods are generally iterative and give an approximate of the exact answer. The accuracy of this approximation depends on the type of method and the stopping step.

Among the five error sources mentioned, the error of the model and the error of the data depend on the type of problem, and the people who determine the model of problem in different disciplines are responsible for them. But the next three errors are related to the numerical analysis.

2.2.2 Round of Error and Floating Points Arithmetic

The problem (2.2.1.6) is in some way, the same as the theorem (2.2.1.7) but is examined from another perspective. Therefore, it is necessary to introduce the definitions and concepts related to floating point. First, we define the floating point.

2.2.2.1 Note

For each computational device, a set of real numbers with finite digits is defined. This set of numbers is called the set of machine numbers and is denoted by the symbol A.

2.2.2.2 Definition

Suppose that $x \notin A$, $rd(x) \in A$ is x approximation if it satisfies the following inequality:

$$\forall g\left(g \in A \Rightarrow |x - rd(x)| \le |x - g|\right)$$

$rd(x)$ can be rounded number to x.

If the computational device has t digits, that is, the number of Mantis significant digits of machine number is maximum of t digits, then $rd(x)$ will be defined as follows:

$$rd(x) = x(1 + \varepsilon), |\varepsilon| \le eps$$

where $eps = 5 \times 10^{-t}$ is called machine accuracy.

2.2.2.3 Definition

The scientific representation of a non-zero number such as binary or decimal x is called the floating points representation. In other words, by changing the exponent, the point shifts among the Mantis digits.

The calculations with these numbers are called floating point calculus.

The result of arithmetic operations on machine numbers may or may not be machine numbers, so it cannot be accepted that arithmetic operations in t-digit computers is accurate. For this purpose, we introduce $+^*$, $-^*$, $\times^*$, and $/^*$ as floating points operations, which are as follows for $x, y \in A$:

$$x +^* y := rd(x + y) = (x + y)(1 + \varepsilon_1)$$

$$x -^* y := rd(x - y) = (x - y)(1 + \varepsilon_2)$$

$$x \times^* y := rd(x \times y) = (x \times y)(1 + \varepsilon_3)$$

$$x /^* y := rd(x / y) = (x / y)(1 + \varepsilon_4)$$

$$|\varepsilon_i| \le eps, \quad i = 1, \ldots 4$$

Therefore, it can be said that floating point arithmetic operations do not satisfy the ordinary arithmetic rules such as having a neutral element, associativity, and distributivity.

2.2.2.4 Note

Suppose E is an algebraic term derived from a floating point calculus. In this case, $fl(E)$ is called the value of the term E in the computational device.

The following theorem shows that the error in the truncating method is almost twice the error in the rounding method.

2.2.2.5 Remark

If $fl(x)$ is the floating points approximation of the non-zero number x in a t-digit computer, then:

$$|x - fl(x)| \le 5 \times |x| \times 10^{-t} \times l$$

where

$$l = \begin{cases} 1 & \text{rounding} \\ 2 & \text{runcating} \end{cases}$$

According to the theorem (2.2.2.5), computational devices use the rounding method to store numbers, in which case:

$$\frac{|x - fl(x)|}{|x|} \le 5 \times 10^{-t} = eps$$

as a result:

$$fl(x) = x(1 + \varepsilon), |\varepsilon| \le eps \qquad (2.7)$$

It can be easily shown that performing floating point operations on numbers will cause the error to propagate and grow, which we will examine in the next section.

Now, we are going to solve some important and applied problems related to the concept of floating point.

2.2.2.6 Problem

Consider the following series:

$$\sum_{i=1}^{n} x_i = x_1 + x_2 + \cdots + x_n$$

Find the floating point approximation of this series.
Solution: Consider:

$$s_1 = x_1$$

$$s_r = fl(s_{r-1} + x_r)$$
$$= (s_{r-1} + x_r)(1 + \varepsilon_r), \quad r = 2, \ldots, n$$

where $|\varepsilon_r| \leq eps$. In this case:

$$s_n = l(s_{n-1} + x_n)(1 + \varepsilon_n)$$

$$= \left[(s_{r-2} + x_{n-1})(1 + \varepsilon_{n-1}) + x_n \right](1 + \varepsilon_n)$$

$$\vdots$$

$$= x_1(1 + \eta_1) + \cdots + x_n(1 + \eta_n)$$

where

$$1 + \eta_r = (1 + \varepsilon_r)(1 + \varepsilon_{r+1})\ldots(1 + \varepsilon_n), |\varepsilon_r| \leq eps, r = 2, \ldots, n$$

$$\eta_1 = \eta_2$$

So,

$$(1 - eps)^{n-r+1} \leq 1 + \eta_r \leq (1 + eps)^{n-r+1}$$

As a result,

$$fl\left(\sum_{i=1}^{n} x_i \right) = \left(\sum_{i=1}^{n} x_i \right)\left(1 + \frac{\sum\limits_{i=1}^{n} x_i \eta_i}{\sum\limits_{i=1}^{n} x_i} \right)$$

2.2.2.7 Problem

Using the t-digit floating point calculus shows

$$rd(a) = \frac{a}{1 + \varepsilon}, |\varepsilon| \leq 5 \times 10^{-t}$$

Solution: We know:

$$rd(a) = a(1 + \varepsilon), |\varepsilon| \leq 5 \times 10^{-t}$$

δ can be defined such that it satisfies the property of ε. For this purpose, it is sufficient to consider δ such that $|\delta| \leq 5 \times 10^{-t}$.

We put $\delta = \dfrac{-\varepsilon}{1 + \varepsilon}$, then:

$$rd(a) = a(1 + \delta), |\delta| \leq 5 \times 10^{-t}$$

$$= a\left(1 + \frac{-\varepsilon}{1 + \varepsilon} \right), |\varepsilon| \leq 5 \times 10^{-t} = a\left(\frac{1}{1 + \varepsilon} \right)$$

2.2.2.8 Problem

Suppose that x is a floating point machine number on a computer with a rounding error unit of ε. It shows

$$fl(x^k) = x^k(1+\delta)^{k+1}, |\delta| \le \varepsilon$$

Solution: We prove it by induction on k.
  Initial case: Suppose $k = 1$. We have:

$$fl(x) = x(1+\delta)^0 = x$$

Induction hypothesis: Suppose that the statement holds for $k = n > 1$.
  Proof of induction hypothesis:
  We prove that the statement also holds for $k = n+1$

$$fl(x^{n+1}) = fl(x^n \cdot x)$$

$$= fl(fl(x^n) \cdot x)$$

$$= fl(x^n) \cdot x \cdot (1+\delta)$$

$$= x^n \cdot (1+\delta)^{n-1} \cdot x \cdot (1+\delta)$$

$$= x^{n+1} \cdot (1+\delta)^n$$

2.2.2.9 Problem

Suppose $|fl(ab) - ab| \le |ab|\beta^{-t+1}$. Calculate the upper bound of the following equation:

$$fl(fl(ab).c) - abc$$

Solution:

$$\left|fl(fl(ab) - abc)\right| \le \left|fl(fl(ab) \cdot c) - fl(ab) \cdot c\right| + \left|fl(ab) \cdot c - abc\right|$$

$$\le |fl(ab) \cdot c|\beta^{-t+1} + |abc|\beta^{-t+1}$$

$$\le \left(|fl(ab) \cdot c - abc| + |abc|\right)\beta^{-t+1} + |abc|\beta^{-t+1}$$

$$\le |abc|\left(1 + \beta^{-t+1}\right)\beta^{-t+1} + |abc|\beta^{-t+1}$$

$$\le |abc|\left(2 + \beta^{-t+1}\right)\beta^{-t+1}$$

2.2.2.10 Problem

Suppose $S_n = \displaystyle\sum_{i=1}^{n} x_i$ where every x_i is a machine number and suppose that S_n^* is what the computer calculates. In this case, $S_n^* = fl(S_{n-1}^* + x_n)$. Prove:

$$S_n^* \simeq S_n + S_2\delta_2 + S_3\delta_3 + \cdots + S_n\delta_n, |\delta_k| \leq eps$$

Also show:

$$S_n^* - S_n \simeq x_1(\delta_2 + \cdots + \delta_n) + x_2(\delta_2 + \cdots + \delta_n) + x_3(\delta_3 + \cdots + \delta_n) + \cdots + x_n\delta_n$$

Solution: We put:

$$S_2^* = fl(x_1 + x_2) = (x_1 + x_2)(1 + \delta_2), |\delta_2| \leq eps$$

And we get by iteration:

$$S_{r+1}^* = fl(S_r^* + x_{r+1}) = (S_r^* + x_{r+1})(1 + \delta_{r+1}), |\delta_{r+1}| \leq eps, r = 1, \ldots, n-1$$

So, we will have:

$$S_2^* - (x_1 + x_2) = \delta_2(x_1 + x_2)$$

$$S_3^* - (x_1 + x_2 + x_3) = fl(S_2^* + x_3) - (x_1 + x_2 + x_3)$$
$$= ((x_1 + x_2)(1 + \delta_2) + x_3)(1 + \delta_3) - (x_1 + x_2 + x_3)$$
$$= \delta_2(x_1 + x_2) + \delta_3(x_1 + x_2 + x_3) + \delta_2\delta_3(x_1 + x_2)$$
$$\simeq \delta_2(x_1 + x_2) + \delta_3(x_1 + x_2 + x_3)$$

Finally, it can be concluded using induction that:

$$S_n^* \simeq S_n + \delta_2(x_1 + x_2) + \delta_3(x_1 + x_2 + x_3) + \cdots + \delta_n(x_1 + \cdots + x_n)$$
$$= S_n + S_2\delta_2 + \cdots + S_n\delta_n$$

Also

$$S_n^* - S_n \simeq \delta_2(x_1 + x_2) + \delta_3(x_1 + x_2 + x_3) + \cdots + \delta_n(x_1 + \cdots + x_n)$$
$$= x_1(\delta_2 + \cdots + \delta_n) + x_2(\delta_2 + \cdots + \delta_n) + x_3(\delta_3 + \cdots + \delta_n) + \cdots + x_n\delta_n$$

2.2.2.11 Problem

Suppose that $x_0, \ldots, x_n$ are positive machine numbers in a computer with a rounding error of ε. Prove that the relative rounding error in the calculation of $\sum_{i=0}^{n} x_i$, in the ordinary way, is a maximum of $(1+\varepsilon)^n - 1$, which is approximately equal to $n\varepsilon$.

Solution: Suppose that $S_k = x_0 + \cdots + x_k$, and S_k^* is the value that the computer calculates instead of S_k. The iterative formulas for these quantities are:

$$\begin{cases} S_0 = x_0 \\ S_{k+1} = S_k + x_{k+1} \end{cases} \qquad \begin{cases} S_0^* = x_0 \\ S_{k+1}^* = fl\left(S_k^* + x_{k+1}\right) \end{cases}$$

For analysis, we define the following values:

$$\rho_k = \frac{S_k^* - S_k}{S_k}, \; \delta_k = \frac{S_{k+1}^* - \left(S_k^* + x_{k+1}\right)}{S_k^* + x_{k+1}}$$

Therefore, $|\rho_k|$ are relative errors for approximating the k th partial sum of S_k by the calculated partial sum of S_k^* and $|\delta_k|$ are relative errors in the approximation of $S_k^* + x_{k+1}$ by the quantity $fl\left(S_k^* + x_{k+1}\right)$. Using the relationships that define ρ_k and δ_k, we have:

$$\rho_{k+1} = \frac{\left(S_{k+1}^* - S_{k+1}\right)}{S_{k+1}}$$

$$= \frac{\left(S_k^* + x_{k+1}\right)\left(1+\delta_k\right) - \left(S_k + x_{k+1}\right)}{S_{k+1}}$$

$$= \frac{\left[S_k\left(1+\delta_k\right) + x_{k+1}\right]\left(1+\delta_k\right) - \left(S_k + x_{k+1}\right)}{S_{k+1}}$$

$$= \delta_k + \rho_k\left(\frac{S_k}{S_{k+1}}\right)\left(1+\delta_k\right)$$

Because $S_k < S_{k+1}$ and $|\delta_k| \le \varepsilon$, we conclude:

$$|\rho_{k+1}| \le \varepsilon + |\rho_k|(1+\varepsilon) = \varepsilon + \theta|\rho_k|$$

where $\theta = 1+\varepsilon$. So, we have the following consecutive inequalities:

$$|\rho_0| = 0$$

$$|\rho_1| \le \varepsilon$$

$$|\rho_2| \le \varepsilon + \theta\varepsilon$$

$$|\rho_3| \le \varepsilon + \theta(\varepsilon + \theta\varepsilon) = \varepsilon + \theta\varepsilon + \theta^2\varepsilon$$

$$\vdots$$

The final result is:

$$|\rho_n| \le \varepsilon + \theta\varepsilon + \theta^2\varepsilon + \cdots + \theta^{n-1}\varepsilon$$

$$= \varepsilon\left(1 + \theta + \theta^2 + \cdots + \theta^{n-1}\right)$$

$$= \varepsilon \cdot \frac{\theta^n - 1}{\theta - 1}$$

$$= \varepsilon \cdot \frac{(1+\varepsilon)^n - 1}{\varepsilon}$$

$$= (1+\varepsilon)^n - 1$$

According to the binomial theorem, we have:

$$(1+\varepsilon)^n - 1 = 1 + \binom{n}{1}\varepsilon + \binom{n}{2}\varepsilon^2 + \cdots - 1 \approx n\varepsilon$$

2.2.2.12 *Problem*

Suppose that $f(x) = x^{2^n}$ and $x = x_0$ is not machine number. So, on the computer, it will have a modified value of $x = x_0(1+\delta)$. Estimate how much $f(x_0)$ will change?

Solution: It is obvious that the value of function $f(x)$ at point x_0 is calculated using n squares by performing a sequence of calculations as follows:

$$x_1 := x_0^2,\ x_2 := x_1^2, \ldots, x_n := x_{n-1}^2$$

which will eventually be equal to $f(x_0) = x_n$. According to the floating point calculus, we will have:

$$\hat{x}_1 = x_0^2(1+\delta_1),\ \hat{x}_2 = x_1^2(1+\delta_2), \ldots, \hat{x}_n = x_{n-1}^2(1+\delta_n)$$

where for $i = 1,\ldots,n$, $|\delta_i| \le eps$. The nested substitution of relationships can result in:

$$\hat{x}_n = x_0^{2^n} \left(1+\delta_n\right)^{2^{n-1}} \ldots \left(1+\delta_{n-1}\right)^2 \left(1+\delta_n\right)$$

Now, we suppose that $\delta = \max_{1 \le i \le n} \delta_i$, so:

$$\hat{x}_n \approx x_0^{2^n} \left(1+\delta_n\right)^{2^n - 1}$$

We define η such that $|\eta| \le eps$ and

$$(1+\delta)^r \approx (1+\eta)^{r+1}$$

So, we conclude:

$$\hat{x}_n \approx x_0^{2^n} \left(1+\eta\right)^{2^n} = f\left(x_0(1+\eta)\right)$$

2.3 Interval Arithmetic

If machine numbers are A approximate numbers, that is,

$$a \in \mathbb{R}, \quad a \in \tilde{a}$$

where

$$\tilde{a} := \left[\underline{a},\overline{a}\right] = \left\{x \mid x \in \mathbb{R}, \ \underline{a} \le x \le \overline{a}\right\}$$

In this case, we say $\tilde{a}$ is an interval number.

Arithmetic operations on interval numbers are defined as follows.

1. The sum of two intervals

$$\tilde{a} + \tilde{b} = \left[\underline{a},\overline{a}\right] + \left[\underline{b},\overline{b}\right] = \left[\underline{a} + \underline{b}, \overline{a} + \overline{b}\right]$$

2. Symmetry of an interval number

$$-\tilde{a} = -\left[\underline{a},\overline{a}\right] = \left[-\overline{a},-\underline{a}\right]$$

3. The difference between two interval numbers:

$$\tilde{a} - \tilde{b} = \left[\underline{a},\overline{a}\right] - \left[\underline{b},\overline{b}\right] = \left[\underline{a} - \overline{b}, \overline{a} - \underline{b}\right]$$

4. The product of a scalar to an interval number:

$$k\tilde{a} = k[\underline{a},\bar{a}] = \begin{cases} [k\underline{a}, k\bar{a}], & k \geq 0 \\ [k\bar{a}, k\underline{a}], & k < 0 \end{cases}$$

5. The product of two intervals

$$\tilde{a} \cdot \tilde{b} = [\underline{a},\bar{a}] \cdot [\underline{b},\bar{b}] = [\underline{c},\bar{c}]$$

where

$$\underline{c} = \min\{\underline{a} \cdot \underline{b}, \underline{a} \cdot \bar{b}, \bar{a} \cdot \underline{b}, \bar{a} \cdot \bar{b}\}$$
$$\bar{c} = \max\{\underline{a} \cdot \underline{b}, \underline{a} \cdot \bar{b}, \bar{a} \cdot \underline{b}, \bar{a} \cdot \bar{b}\}$$

6. The division of two interval numbers:

$$\frac{\tilde{a}}{\tilde{b}} = \frac{[\underline{a},\bar{a}]}{[\underline{b},\bar{b}]} = [\underline{c},\bar{c}]$$

where

$$\underline{c} = \min\left\{\frac{\underline{a}}{\underline{b}}, \frac{\bar{a}}{\underline{b}}, \frac{\underline{a}}{\bar{b}}, \frac{\bar{a}}{\bar{b}}\right\}$$

$$\bar{c} = \max\left\{\frac{\underline{a}}{\underline{b}}, \frac{\bar{a}}{\underline{b}}, \frac{\underline{a}}{\bar{b}}, \frac{\bar{a}}{\bar{b}}\right\}$$

2.4 Interval Error

Suppose that the interval number $\tilde{a}$ is approximately equal to the interval number $\tilde{A}$. So

$$\tilde{a} \approx \tilde{A} \Leftrightarrow \left(\underline{a} \approx \underline{A} \,\&\, \bar{a} \approx \bar{A}\right)$$

In this case, absolute and relative error of $\tilde{a}$ are defined as follows:

$$\Delta(\tilde{a}) = \tilde{A} - \tilde{a} = [\bar{A}, \underline{A}] - [\underline{a}, \bar{a}] = [\underline{A} - \bar{a}, \bar{A} - \underline{a}]$$

$$\varepsilon_{\tilde{a}} = \frac{\tilde{A} - \tilde{a}}{\tilde{A}} = \frac{\left[\underline{A} - \tilde{a}, \overline{A} - a\right]}{\left[\underline{A}, \overline{A}\right]} = \left[\underline{c}, \overline{c}\right]$$

where

$$\underline{c} = \min\left[\frac{\underline{A} - \tilde{a}}{\underline{A}}, \frac{\underline{A} - \tilde{a}}{\overline{A}}, \frac{\overline{A} - a}{\underline{A}}, \frac{\overline{A} - a}{\overline{A}}\right]$$

$$\underline{c} = \max\left[\frac{\underline{A} - \tilde{a}}{\underline{A}}, \frac{\underline{A} - \tilde{a}}{\overline{A}}, \frac{\overline{A} - a}{\underline{A}}, \frac{\overline{A} - a}{\overline{A}}\right]$$

Therefore, it can be said that the smaller the corresponding absolute and relative error, the smaller the maximum error.

2.5 Interval Floating Point Calculus

Suppose that $\tilde{A}$ as a set of approximate machine numbers is defined as follows:

$$\tilde{A} = \left\{a \mid a = [\underline{a}, \overline{a}]\right\} = \left[\underline{A}, \overline{A}\right]$$

where:

$$\underline{A} = \left\{\underline{a} \mid \tilde{a} \in \tilde{A} \,\&\, \tilde{a} = [\underline{a}, \overline{a}]\right\}$$

$$\overline{A} = \left\{\overline{a} \mid \tilde{a} \in \tilde{A} \,\&\, \tilde{a} = [\underline{a}, \overline{a}]\right\}$$

In this case:

(I) $\tilde{x} \in \tilde{A}$ $\Leftrightarrow$ $\exists \tilde{a}\left(\tilde{a} \in \tilde{A} \,\&\, \tilde{x} = \tilde{a} \,\&\, fl(\tilde{x}) = \tilde{x}\right)$

(II) $\tilde{x} \notin \tilde{A}$ $\Leftrightarrow$ $fl(\tilde{x}) = \tilde{x}(1 + \varepsilon), \quad |\varepsilon| \le eps$

(2.8)

which *eps* is defined on the interval calculus as follows:

$$eps = \min\left\{\tilde{g} \in \tilde{A} \middle| 1 + \overset{*}{} \tilde{g} > 1\right\}$$

(2.9)

where $1 = [1, 1]$.

Given that:

$$\tilde{a} \geq 0 \Rightarrow [\underline{a}, \overline{a}] \geq [0,0] \Rightarrow (\underline{a} \geq 0 \ \& \ \overline{a} \geq 0 \ \& \ \underline{a} \leq \overline{a})$$
$$\tilde{a} \leq \tilde{b} \Leftrightarrow (\underline{a} \leq \underline{b} \ \& \ \overline{a} \leq \overline{b})$$

Therefore, (2.9) can be written as follows:

$$eps = \min\left\{ [\underline{g}, \overline{g}] \in \tilde{A} \Big| 1 + \overset{*}{} \underline{g} > 1 \ \& \ 1 + \overset{*}{} \overline{g} > 1 \right\}$$

Note that since $\tilde{A}$ is finite, it can take its minimum value, so *eps* is equal to the minimum of the above set, not its infimum.

If we define $eps := [\underline{e}, \overline{e}]$, we have:

$$\underline{e} = \min\left\{ \underline{g} \in \underline{A} \Big| 1 + \overset{*}{} \underline{g} > 1 \right\}$$
$$\underline{e} = \min\left\{ \overline{g} \in \overline{A} \Big| 1 + \overset{*}{} \overline{g} > 1 \right\}$$

where $\underline{e}$ is the minimum lower limit of the intervals and $\overline{e}$ is the minimum upper limit of the intervals.

According to the equation (2.8), we have:

$$fl(\tilde{x}) = \tilde{x}(1+\varepsilon)$$
$$= [\underline{x}, \overline{x}](1+\varepsilon)$$
$$= [\underline{x}(1+\varepsilon), \overline{x}(1+\varepsilon)]$$
$$= [fl(\underline{x}), fl(\overline{x})]$$
$$= [\underline{fl}(\tilde{x}), \overline{fl}(\tilde{x})]$$

where $|\varepsilon| \leq eps = \tilde{5} \times 10^{-t} = [\underline{5} \times 10^{-t}, \overline{5} \times 10^{-t}] = [\underline{e}, \overline{e}]$.

2.6 Problem

Find the floating point error of the arithmetic operations on two arbitrary intervals and then generalize it to *n* intervals.

Solution: Suppose that $\tilde{a} = [\underline{a}, \overline{a}]$ and $\tilde{b} = [\underline{b}, \overline{b}]$ are two arbitrary intervals. For the sum of two intervals, we have:

$$\tilde{a} +^{*} \tilde{b} = fl\left(\tilde{a} + \tilde{b}\right)$$

$$= \left(\left[\underline{a}, \overline{a}\right] + \left[\underline{b}, \overline{b}\right]\right)(1 + \varepsilon)$$

$$= \left[\left(\underline{a} + \underline{b}\right)(1 + \varepsilon), \left(\overline{a} + \overline{b}\right)(1 + \varepsilon)\right], \quad |\varepsilon| \le eps$$

Therefore, for the sum of n intervals, we can write:

$$\sum_{i=1}^{n} {}^{*}\tilde{a}_i = fl\left(\sum_{i=1}^{n} \tilde{a}_i\right) = \left[\sum_{i=1}^{n} \underline{a}_i(1 + \varepsilon), \sum_{i=1}^{n} \overline{a}_i(1 + \varepsilon)\right], \quad |\varepsilon| \le eps$$

For the difference between two intervals, we have:

$$\tilde{a} -^{*} \tilde{b} = fl\left(\tilde{a} - \tilde{b}\right)$$

$$= \left(\tilde{a} - \tilde{b}\right)(1 + \varepsilon)$$

$$= \left[\left(\underline{a} - \overline{b}\right)(1 + \varepsilon), \left(\overline{a} - \underline{b}\right)(1 + \varepsilon)\right], \quad |\varepsilon| \le eps$$

So, for the difference of n intervals, we will have:

$$\left[\underline{a}_1, \overline{a}_1\right] -^{*} \left[\underline{a}_2, \overline{a}_2\right] -^{*} \cdots -^{*} \left[\underline{a}_n, \overline{a}_n\right]$$

$$= \left[\underline{a}_1 -^{*} \sum_{i=2}^{n} {}^{*}\overline{a}_i, \overline{a}_1 -^{*} \sum_{i=2}^{n} {}^{*}\underline{a}_j\right]$$

$$= fl\left[\underline{a}_1 - \sum_{i=2}^{n} \overline{a}_i, \overline{a}_1 - \sum_{i=2}^{n} \underline{a}_j\right]$$

$$= \left[\left(\underline{a}_1 - fl\left(\sum_{i=2}^{n} \overline{a}_i\right)\right)(1 + \delta), \left(\overline{a}_1 - fl\left(\sum_{i=2}^{n} \underline{a}_j\right)\right)(1 + \delta)\right]$$

$$= \left[\left(\underline{a}_1 - \left(\sum_{i=2}^{n} \overline{a}_i\right)(1 + \varepsilon)\right)(1 + \delta), \left(\overline{a}_1 - l\left(\sum_{i=2}^{n} \underline{a}_j\right)(1 + \varepsilon)\right)(1 + \delta)\right]$$

where $|\delta| \le eps$ and $|\varepsilon| \le eps$,

For the product of two intervals, we have:

$$\tilde{a} \cdot^* \tilde{b} = fl\left(\tilde{a} \cdot \tilde{b}\right)$$

$$= fl\left(\left[\underline{a}, \overline{a}\right] \cdot \left[\underline{b}, \overline{b}\right]\right)$$

$$= \left(\left[\underline{a}, \overline{a}\right] \cdot \left[\underline{b}, \overline{b}\right]\right)(1+\varepsilon)$$

$$= \left(\left[\underline{c}, \overline{c}\right]\right)(1+\varepsilon)$$

where

$$\underline{c} = \min\left\{\underline{a} \cdot \underline{b}, \underline{a} \cdot \overline{b}, \overline{a} \cdot \underline{b}, \overline{a} \cdot \overline{b}\right\}$$

$$\overline{c} = \max\left\{\underline{a} \cdot \underline{b}, \underline{a} \cdot \overline{b}, \overline{a} \cdot \underline{b}, \overline{a} \cdot \overline{b}\right\}$$

$$|\varepsilon| \leq eps$$

Now for the product of n intervals, we have:

$$p_1(\tilde{a}_1) = \left\{\underline{a}_1, \overline{a}_1\right\}(1+\varepsilon_1)$$

$$I_{p_1} = \left[\min p_1, \max p_1\right]$$

$$p_2(\tilde{a}_1, \tilde{a}_2) = \left\{\underline{a}_1 \cdot \underline{a}_2, \underline{a}_1 \cdot \overline{a}_2, \overline{a}_1 \cdot \underline{a}_2, \overline{a}_1 \cdot \overline{a}_2\right\}(1+\varepsilon_1)$$

$$I_{p_2} = \left[\min p_2, \max p_2\right]$$

$$p_3(\tilde{a}_1 \cdot \tilde{a}_2 \cdot \tilde{a}_3) = \left\{\min p_2 \cdot \underline{a}_3, \min p_2 \cdot \overline{a}_3, \max p_2 \cdot \underline{a}_3, \max p_2 \cdot \overline{a}_3\right\}(1+\varepsilon_1)$$

$$I_{p_3} = \left[\min p_3, \max p_3\right]$$

Therefore, it can be written:

$$p_n\left(\prod_{i=1}^{n} \tilde{a}_i\right) = \left\{\min p_{n-1} \cdot \underline{a}_n, \min p_{n-1} \cdot \overline{a}_n, \max p_{n-1} \cdot \underline{a}_n, \max p_{n-1} \cdot \overline{a}_n\right\}(1+\varepsilon_n)$$

$$I_{p_n} = \left[\min p_n, \max p_n\right]$$

2.7 Algorithm Error Propagation

In this section, we examine the propagation of rounding error in an algorithm. As mentioned earlier,

$$\phi = \varphi^{(r)} \ldots \varphi^{(0)}$$

If we define $X^{(0)} = X$, then Y is obtained as follows:

$$X = X^{(0)} \rightarrow \varphi^{(0)}\left(X^{(0)}\right) \rightarrow \cdots \rightarrow \varphi^{(r)}\left(X^{(r)}\right) = X^{(r+1)} = Y \qquad (2.10)$$

Suppose that we have the map $\psi^{(i)}$ (also called the remaining map) as follows:

$$\psi^{(i)} = \varphi^{(r)} \ldots \varphi^{(0)}, \psi^{(i)} : D_i \rightarrow \mathbb{R}^m, i = 0, \ldots, r$$

where $\psi^{(0)} \equiv \phi$, which means that none of algorithm steps have been performed. So for $i = 0, \ldots, r$, we have:

$$D\psi^{(i)}\left(X^{(i)}\right) = D\varphi^{(r)}\left(X^{(r)}\right) \cdot D\varphi^{(r-1)}\left(X^{(r-1)}\right) \ldots D\varphi^{(i)}\left(X^{(i)}\right) \qquad (2.11)$$

where $D\psi^{(i)}$ is a Jacobian matrix of map $\psi^{(i)}$. Therefore, the propagation of rounding error in the steps of performing an algorithm can be defined as follows:

$$\Delta\tilde{X}^{(1)} \simeq D\varphi^{(0)}(X) \cdot \Delta\tilde{X} + \alpha_1$$

$$\Delta\tilde{X}^{(2)} \simeq D\varphi^{(1)}\left(X^{(1)}\right)\left(D\varphi^{(0)}(X) \cdot \Delta\tilde{X} + \alpha_1\right) + \alpha_2$$

$$\vdots$$

$$\Delta\tilde{X}^{(r+1)} = \Delta\tilde{Y}$$

$$\simeq D\varphi^{(r)}\left(X^{(r)}\right) \ldots D\varphi^{(0)}(X) \cdot \Delta\tilde{X} + D\varphi^{(r)}\left(X^{(r)}\right) \ldots D\varphi^{(1)}\left(X^{(1)}\right) \cdot \alpha_1$$

$$+ \cdots + D\varphi^{(r)}\left(X^{(r)}\right) \cdot \alpha_r + \alpha_{r+1}$$

Now if for $i = 0, \ldots, r$, we define $\alpha_{i+1} := E_{i+1} \cdot X^{(i+1)}$, where

$$E_{i+1} := \begin{bmatrix} \varepsilon_1 & \cdots & 0 \\ \vdots & \ddots & \vdots \\ 0 & \cdots & \varepsilon_{n_i+1} \end{bmatrix}$$

is the error diagonal matrix, we will have:

$$\Delta \tilde{Y} = D\phi(X) \cdot \Delta \tilde{X} + D\psi^{(1)}\left(X^{(1)}\right) \cdot E_1 \cdot X^{(1)} + \cdots + D\psi^{(r)}\left(X^{(r)}\right) \cdot E_r \cdot X^{(r)} + E_{r+1} \cdot Y$$

$$(2.12)$$

Now we suppose that no step of the algorithm is executed, so in the equation (2.12), we will have only $D\phi(X) \cdot \Delta \tilde{X}$ and $E_{r+1} \cdot Y$. So $\Delta \tilde{Y}$ will be as follows:

$$\Delta^{(0)} \tilde{Y} = D\phi(X) \cdot \Delta \tilde{X} + + E_{r+1} \cdot Y$$

$$= \left(\left|D\phi(X)\right| \cdot |X| + |Y|\right) eps$$

which is called the inherent error of the algorithm. It is clear that the inherent error depends on the input vector of X and the output vector of Y and will appear in any case.

If for every i, the rounding error propagations (E_i's) on the total error are such that:

$$\left|D\psi^{(i)}\left(X^{(i)}\right) \cdot E_i \cdot X^{(i)}\right| \simeq \Delta^{(0)} \tilde{Y}$$

Then, we say the rounding errors are harmless. If the rounding errors are harmless at all steps, the algorithm is well established or numerically stable.

2.7.1 Problem

The following function is calculated for $0 \le \theta \le 2\pi$ and $0 < k_c \le 1$:

$$f(\theta, k_c) := \frac{1}{\sqrt{\cos^2 \theta + k_c^2 \sin^2 \theta}}$$

If we apply the following method:

$$k^2 := 1 - k_c^2$$

$$f(\theta, k_c) := \frac{1}{\sqrt{1 - + k^2 \sin^2 \theta}}$$

In this case, we do not need to calculate $\cos \theta$ and the method is faster. Compare both methods for numerical stability, as follows:

Solution: Algorithm 1:

$$\phi(a, b) = \cos^2 a + b \sin^2 a$$

If $x = x^{(0)} = \begin{bmatrix} a \\ b \end{bmatrix}$, then:

$$\varphi^{(0)}\left(x^{(0)}\right) = \begin{bmatrix} \cos a \\ \sin a \\ b \end{bmatrix} = x^{(1)}$$

$$\varphi^{(1)}\left(x^{(1)}\right) = \begin{bmatrix} \cos^2 a \\ \sin^2 a \\ b \end{bmatrix} = x^{(2)}$$

$$\varphi^{(2)}\left(x^{(2)}\right) = \begin{bmatrix} \cos^2 a \\ b\sin^2 a \end{bmatrix} = x^{(3)}$$

$$\varphi^{(3)}\left(x^{(3)}\right) = \cos^2 a + b\sin^2 a = x^{(4)} = \phi(a,b)$$

So, we have:

$$\psi^{(1)}\left(u,v,w\right) = \varphi^{(3)} \circ \varphi^{(2)} \circ \varphi^{(1)}\left(u,v,w\right) = u^2 + wv^2$$

Then:

$$D\psi^{(1)}\left(u,v,w\right) = \left(2u, 2wv, v^2\right)$$

By substituting $x^{(1)}$, we have:

$$D\psi^{(1)}\left(x^{(1)}\right) = \left(2\cos a, 2b\sin a, \sin^2 a\right)$$

and also:

$$\psi^{(2)}\left(r,s,t\right) = \varphi^{(3)} \circ \varphi^{(2)}\left(r,s,t\right) = r + st$$

So:

$$D\psi^{(2)}\left(r,s,t\right) = \left(1, s, t\right)$$

By considering $x^{(2)}$, we have:

$$D\psi^{(2)}\left(x^{(2)}\right) = \left(1, b, \sin^2 a\right)$$

and also:

$$\psi^{(3)}\left(m,n\right) = m + n$$

So:

$$D\psi^{(3)}(m,n) = (1,1)$$

By substituting $x^{(3)}$, you can write:

$$D\psi^{(3)}\left(x^{(3)}\right) = (1,1)$$

and finally:

$$D\phi(x) = D\phi(a,b) = \left(-2\sin a\cos a + 2b\sin a\cos a, \sin^2 a\right)$$

and according to the above relations, it can be written:

$$\alpha_1 = \begin{bmatrix} \varepsilon_1\cos a \\ \varepsilon_2\sin a \\ 0 \end{bmatrix}, \alpha_2 = \begin{bmatrix} \varepsilon_3\cos^2 a \\ \varepsilon_4\sin^2 a \\ 0 \end{bmatrix}$$

$$\alpha_3 = \begin{bmatrix} 0 \\ \varepsilon_5 b\sin^2 a \end{bmatrix}, \alpha_4 = \varepsilon_6\left(\cos^2 a + b\sin^2 a\right)$$

Since:

$$\Delta\tilde{x} = |\tilde{x} - x| = |fl(x) - x| = |x(1+\varepsilon) - x| = |x|\varepsilon, |\varepsilon| \le eps \tag{2.13}$$

finally, we will have:

$$\Delta\tilde{y} = D\phi(x)\Delta\tilde{x} + D\psi^{(1)}\left(x^{(1)}\right)\alpha_1 + D\psi^{(2)}\left(x^{(2)}\right)\alpha_2 + D\psi^{(3)}\left(x^{(3)}\right)\alpha_3 + \alpha_4$$

$$= 2(b-1)\sin a\cos a\cdot\Delta a + \sin^2 a\cdot\Delta b + 2\varepsilon_1\cos^2 a$$

$$+2\varepsilon_2\sin^2 a + \varepsilon_3\cos^2 a + b\varepsilon_4\sin^2 a$$

$$+b\varepsilon_5\sin^2 a + \varepsilon_6\cos^2 a + b\varepsilon_6\sin^2 a$$

$$\le 2(b-1)\sin a\cos aa|\varepsilon| + \sin^2 ab|\varepsilon|$$

$$+2\cos^2 a|\varepsilon_1| + 2b\sin^2 a|\varepsilon_2| + \cos^2 a|\varepsilon_3|$$

$$+2b\sin^2 a|\varepsilon_4| + 2b\sin^2 a|\varepsilon_5| + \cos^2 a|\varepsilon_6| + 2b\sin^2 a|\varepsilon_6|$$

$$\le 3eps + \left(4\cos^2 a + 5b\sin^2 a\right)\cdot eps$$

But about algorithm 2:

If $x = x^{(0)} = \begin{bmatrix} a \\ b \end{bmatrix}$, then:

$$\varphi^{(0)}\left(x^{(0)}\right) = \begin{bmatrix} \sin a \\ 1-b \end{bmatrix} = x^{(1)}$$

$$\varphi^{(1)}\left(x^{(1)}\right) = \begin{bmatrix} \sin^2 a \\ 1-b \end{bmatrix} = x^{(2)}$$

$$\varphi^{(2)}\left(x^{(2)}\right) = (1-b)\sin^2 a = x^{(3)}$$

$$\varphi^{(3)}\left(x^{(3)}\right) = 1-(1-b)\sin^2 a = x^{(4)} = \phi(a,b)$$

So:

$$\psi^{(1)}(m,n) = \varphi^{(3)} \circ \varphi^{(2)} \circ \varphi^{(1)}(m,n) = 1 - m^2 n$$

Then:

$$D\psi^{(1)}(m,n) = \left(-2mn, -m^2\right)$$

By substituting $x^{(1)}$ we have:

$$D\psi^{(1)}\left(x^{(1)}\right) = \left(-2(1-b)\sin a, -\sin^2 a\right)$$

and also:

$$\psi^{(2)}(r,s) = \varphi^{(3)} \circ \varphi^{(2)}(r,s) = 1 - rs$$

So:

$$D\psi^{(2)}(r,s) = (-s, -r)$$

By substituting $x^{(2)}$, we have:

$$D\psi^{(2)}\left(x^{(2)}\right) = \left(-(1-b), -\sin^2 a\right)$$

and also:

$$\psi^{(3)}(t) = 1 - t$$

So:

$$Dψ^{(3)}(t) = -1$$

By substituting $x^{(3)}$, you can write:

$$Dψ^{(3)}\left(x^{(3)}\right) = -1$$

and finally:

$$Dφ(x) = Dφ(a,b) = \left(-2(1-b)\sin a \cos a, \sin^2 a\right)$$

and according to the above relations, it can be written:

$$α_1 = \begin{bmatrix} ε_1 \sin a \\ ε_2(1-b) \end{bmatrix}, α_2 = \begin{bmatrix} ε_3 \sin^2 a \\ 0 \end{bmatrix}$$

$$α_3 = ε_4(1-b)\sin^2 a, α_4 = ε_5\left(1-(1-b)\sin^2 a\right)$$

Given the relation (2.13), we will finally have:

$$Δ\tilde{y} = Dφ(x)Δ\tilde{x} + Dψ^{(1)}\left(x^{(1)}\right)α_1 + Dψ^{(2)}\left(x^{(2)}\right)α_2 + Dψ^{(3)}\left(x^{(3)}\right)α_3 + α_4$$

$$= 2(b-1)\sin a \cos a \cdot Δa + \sin^2 a \cdot Δb$$

$$+2(b-1)ε_1 \sin^2 a + (b-1)ε_2 \sin^2 a + (b-1)ε_3 \sin^2 a$$

$$+(b-1)ε_4 \sin^2 a + ε_5\left(1-(1-b)\sin^2 a\right)$$

$$≤ 2(b-1)\sin a \cos aa|ε| + \sin^2 ab|ε|$$

$$+\left(2(1-b)\sin^2 a + (1-b)\sin^2 a + (1-b)\sin^2 a + ...\right) \cdot eps$$

$$= 3eps + \left(6(1-b)\sin^2 a + 1\right) \cdot eps$$

For the algorithm 2 to be numerically better than the algorithm 1, we must have:

$$6(1-b)\sin^2 a + 1 ≤ 4\cos^2 a + 5b\sin^2 a$$

$$⇒ 6(1-b)\sin^2 a + 1 - 4\cos^2 a - 5b\sin^2 a ≤ 0$$

$$⇒ \sin^2 a(6 - 6b - 5b) - 4\cos^2 a + 1 ≤ 0$$

Given that $0 \le a \le \dfrac{\pi}{2}$, so $\sin a \le 1$ and $\cos a \le 1$. So:

$$6 - 11b \le 3 \Rightarrow b \ge \frac{3}{11}$$

So, if $\dfrac{3}{11} < b < 1$, the algorithm 2 is better than the algorithm 1, otherwise the algorithm 1 would be better, that is, if $0 < b < \dfrac{3}{11}$, the algorithm 1 is better. If $b = \dfrac{3}{11}$ and $0 \le a \le \dfrac{\pi}{2}$, the algorithm 2 still works better than the algorithm 1.

2.7.2 Scientific Representation of Numbers

Suppose that x is a non-zero number. It is obvious that x can always be written as

$$x = p \cdot 10^q, \ 1 \le |p| < 10$$

where y is an integer.

In this case, we can state that x is scientifically represented. In this display, p is called the mantissa and q is called the exponent of the number x.

2.7.3 Definition

If p is a decimal number and $1 \le |p| < 10$, then the significant figures of p are the non-zero figures of p, the zeros between these figures and the zeros in front of the number to indicate the accuracy. The significant figures of the non-zero number x are same as the significant figures of the mantissa of x.

2.7.4 Example

A. If $x = 301.57$, then $x = 3.0157 \times 10^2$ and the number of significant figures of x is 5.

B. If $x = 0.000497$, then $x = 4.97 \times 10^{-4}$; and x has three significant figures.

C. If $x = 2000$ m, then $x = 2.000 \times 10^3$ m; and x has 4 significant figures.

and especially since always $\dfrac{n+1}{n} > 1$, none of the above numbers is equal to the sequence limit. In numerical analysis, we are always faced with such a situation, that is, a sequence of numbers is created that converges to the answer of the considered problem under certain conditions. Thus, we should generally define the error of an arbitrary approximation of a number. For this purpose, we assume that x is a number (accurate) and $\tilde{x}$ is an approximation of that.

2.8 Exercises

1. Suppose that $A \in \mathbb{R}^{n \times n}$ and $X \in \mathbb{R}^n$ and $fl(AX)$ is the result of the inner product of the matrix A and the vector x in the computer. We define:

$$|A| = \left[|a_{ij}| \right]_{i,j} = 1,\ldots,n$$

$$|X| = \left[|x_1|,\ldots,|x_n| \right]^T$$

According to the floating point calculus, $fl(AX) = AX + \varepsilon$ where $e \in \mathbb{R}^n$ is the error vector. Prove that:

$$|e| \leq 1.01 nu |A| \|X\|, \quad 0 \leq nu \leq 0.01$$

2. Suppose that $a, b \in \mathbb{R}$ and $x = a - b$ is defined. The approximations of the floating points of a and b are $\hat{a} = fl(a) = a(1 + \varepsilon_a)$ and $\hat{b} = fl(b) = b(1 + \varepsilon_a)$, respectively. Therefore, the approximation of the floating point x will be $\hat{x} = \hat{a} - \hat{b}$. Show that the error relation is as follows:

$$|\varepsilon| = \left| \frac{x - \hat{x}}{x} \right| \leq \alpha \frac{|a| + |b|}{|a - b|}$$

What is the value of α? Under what circumstances is $|\varepsilon|$ large?

3. For $a \neq 0$, the quadratic equation of $ax^2 + bx + c = 0$ has roots as

$$x_1 = \frac{-b + \sqrt{b^2 - 4ac}}{2a}, \quad x_2 = \frac{-b - \sqrt{b^2 - 4ac}}{2a}$$

and for $c \neq 0$, the quadratic equation of $cx^2 + bx + a = 0$ has roots as

$$x_1' = \frac{-b + \sqrt{b^2 - 4ac}}{2c}, \quad x_2' = \frac{-b - \sqrt{b^2 - 4ac}}{2c}$$

a. Show that $x_1 x_2' = 1$ and $x_2 x_1' = 1$.
b. Using the result of the exercise (2), explain that for $b^2 \gg |4ac|$, which one of x_1 and x_2 is calculated with the better accuracy. Can the result of part a be used to reduce the problem? Explain.

4. Suppose that $f(x)=\sqrt{1+\sin^2 x}-1$ where $|x|$ is a small number. Calculate the condition number for the above function. How to improve the calculation of $f(x)$?

5. Suppose that $x_1,\ldots,x_n$ are positive machine numbers and $S_n = \sum_{i=1}^{n} x_i$ and S_n^* are its corresponding sum in the computer. Prove that if for every i, $x_{i+1} \geq S_i eps$, then

$$\frac{|S_n^* - S_n|}{S_n} \leq (n-1)eps$$

6. Indicate that which of the following statements is not necessarily true? (x, y and z are machine numbers and $|\delta| \leq eps$).

   a. $fl(xy)=xy(1+\delta)$

   b. $fl(x+y)=(x+y)(1+\delta)$

   c. $fl(xy)=\dfrac{xy}{1+\delta}$

   d. $|fl(xy)-xy| \leq |xy| eps$

   e. $fl(x+y+z)=(x+y+z)(1+\delta)$

7. In the sum of n machine numbers, what can be said about the relative rounding error? (These numbers are not necessarily positive).

8. show that

$$fl(x)=\frac{x}{1+\delta}, \quad |\delta| \leq eps$$

9. Prove that if x and y are machine numbers and $|y| \leq |x| eps \times 10^{-1}$, then:

$$fl(x+y)=x$$

10. The following equations both compute variance of S^2 for observations of $x_1,\ldots,x_n$. Which are numerically more reliable?

$$S^2 = \frac{1}{n-1}\left(\sum_{i=1}^{n} x_i^2 - n\bar{x}^2\right)$$

$$S^2 = \frac{1}{n-1} \sum_{i=1}^{n} (x_i - \bar{x})^2$$

where

$$\bar{x} = \frac{1}{n} \sum_{i=1}^{n} x_i$$

11. Calculate the error propagation for the following equation and then state the problems of the y calculation method from the above formula.

$$y = \frac{-b + \sqrt{b^2 - 4ac}}{2a}$$

12. Suppose that $A = a + \varepsilon_a$ and $B = b + \varepsilon_b$. Prove that

$$\Delta\left(\frac{a}{b}\right) \leq \frac{|b| \Delta(a) + |a| \Delta(b)}{|b|^2}$$

3

Interpolation

3.1 Introduction

Considering the many applications of mathematical functions in different methods, it is important to have a function rule. Sometimes, we may only have some data from a function, and if we want to know the behavior of the function, we must have a graph of the function. For example, suppose that a surveying team takes coordinates from a long path with surveying cameras. For instance, this path could be an airport runway or a tunnel between cities or a subway line between city stations. The coordinates obtained from this survey are our data or assumptions. If we know the behavior of these coordinates, that is, if we have a curve connecting these points, then we will have the same airport runway or tunnel or subway line. We know that the aircraft applies a strong impulse to the runway when it lands on the runway and then starts to stop at a high speed of about 250 kilometers/hour. Obviously, if the runway has small bumps, it will shorten the life of the aircraft. Therefore, it is necessary for the runway to be a very smooth path with the least curvature. For this purpose, the function that approximates the behavior of these coordinates must be a smooth, low-curvature function. This can be done using a very powerful interpolation function such as spline. Sometimes, as we consider the approximate location of points or coordinates, the drawn curve may not pass through some points, in which case, we will have another form of approximation called the curve fitting. In both cases, for better approximation, we select an interpolation function that behaves similar to the approximate behavior of the points and their distribution. Recently, the topic of interpolation, especially the splines, has been introduced in medical sciences. For example, it is used in very fine devices that take films or photos from the inside of the intestine and stomach. Other applications of this type of interpolation function are in various engineering sciences, of which everyone is aware.

A general but brief introduction to interpolation functions is given below.

Assume that the values of the function f are $f_0, \ldots, f_n$, respectively, for the mutual distinct points of $x_0, \ldots, x_n$. We call such a function a tabular function. Estimating the value of $f(x)$ when $x \in [x_0, x_n]$ and $x \neq x_i, i = 0, \ldots, n$ is called

---

DOI: 10.1201/9781003218173-3

interpolation and estimating the value of $f(x)$ when $x \in [x_0, x_n]$ is called extrapolation.

Estimating the value of a function for a point like x that is not in the table but between table points is also an interpolation.

Suppose that ϕ is a family of single-variable functions of x with $n+1$ parameters of $a_0, \ldots, a_n$. The problem of interpolation ϕ is to determine the parameters $a_0, \ldots, a_n$ in $\phi(x; a_0, \ldots, a_n)$ so that for pairs $(x_i, f_i), i = 0, \ldots, n$ we have:

$$\phi(x_i; a_0, \ldots, a_n) = f_i, \quad i = 0, \ldots, n \tag{3.1}$$

where the mutual distinct points of $(x_i, f_i), i = 0, \ldots, n$ can be real or complex. The equation (3.1) is called the interpolation problem and points (x_i, f_i) are called the interpolation points.

Obviously, the goal is to determine the $n+1$ unknown parameter of $a_0, \ldots, a_n$ using the set of equations (3.1). If ϕ is linearly dependent on parameters $a_0, \ldots, a_n$, then the interpolation problem (3.1) is called linear, that is,

$$\phi(x; a_0, \ldots, a_n) = a_0 \phi_0(x) + \cdots + a_n \phi_n(x) \tag{3.2}$$

Otherwise, call it nonlinear. For example:
Linear:

1. If $\phi_j(x) = x^j, j = 0, \ldots, n$, then $\phi(x; a_0, \ldots, a_n)$ defines an interpolator as a polynomial of at most degree n.
2. If $\phi_j(x) = e^{ijx}$ or $\phi_j(x) = \cos jx$ or $\phi_j(x) = \sin jx, j = 0, \ldots, n$, then $\phi(x; a_0, \ldots, a_n)$ is a triangular interpolator.
3. If $\phi \in c^2[x_0, x_n]$, then $\phi|_{[x_j, x_{j+1}]} \in \Pi_3$ is a cubic spline interpolator.

Nonlinear:

4. If $\phi(x; a_0, \ldots, a_\mu, b_0, \ldots, b_v) = \dfrac{a_0 + a_1 x + \cdots + a_\mu x^\mu}{b_0 + b_1 x + \cdots + b_v x^v}$, ϕ is a fractional interpolator.

5. If $\phi(x; a_0, \ldots, a_\mu, \lambda_0, \ldots, \lambda_\mu) = a_0 e^{\lambda_0 x} + \cdots + a_\mu e^{\lambda_\mu x}$, then ϕ is an exponential interpolator.

It is worth noting that for each of interpolations that have been proposed, the introduced interpolation of ψ is a linear combination of a set of functions of $\{\phi_0, \ldots, \phi_n\}$, that is, they are generated by its members. These types of functions are defined so that they form an independent linear set. Therefore, it can be claimed that all interpolator functions form a vector space that can have different bases, and each of the interpolators, according to the structure of the basic members, belongs to their own space. Obviously, the representation of each interpolation function by the basic members in the corresponding space is unique, so generally, in the equation (3.2), the linear combination coefficients of a_i, $i = 0,...,n$ are obtained uniquely. Therefore, it can be concluded that any kind of interpolator is unique in its own space. If we want to have a kind of classification in the order of the appearance of the interpolation functions, we can say:

A. Non-recursive interpolations
For example: Lagrange, trigonometric, fractional, and exponential interpolators.

B. Recursive interpolations
For example: Neville, Aitken, and Newton interpolators.

C. Interpolation with multi-criteria functions
For example: Hermit, spline, and B-spline interpolators.

We will now concisely discuss each of these interpolations.

3.2 Lagrange Interpolation

This type of interpolation is a special case of type (1) interpolation.
Polynomial function of

$$p(x) = \sum_{j=0}^{n} L_j(x) f_j \tag{3.3}$$

is a Lagrange interpolator polynomial if it holds in the interpolation condition of $p(x_i) = f_i$, $i = 0,1,...,n$, where

$$L_j(x) = \prod_{i=0, j \neq i}^{n} \frac{x - x_i}{x_j - x_i}, \quad j = 0,...,n \tag{3.4}$$

are Lagrange polynomials of degree n with condition of $L_j(x_k) = \delta_{jk}$, $0 \leq j, k \leq n$. The calculations in the Lagrange method, even if n is not too large, are huge and time-consuming, and automation of the operation is not easy. Also, the degree of interpolation polynomials is determined after the calculations are completed. By adding a point to the table, all operations must be repeated, and since interpolation polynomials are not calculated gradually, this method must be used with caution. Now, we examine some properties of this interpolation using some problems.

3.2.1 Problem

Prove that

$$\sum_{i=0}^{n} L_i(x) = 1$$

Solution: Suppose that $f(x) = 1$, and the interpolation polynomials $p(x)$ for points of $x_0, \ldots, x_n$ are such that $p \in \Pi_n$ and for $i = 0, \ldots, n$, $p(x_i) = 1$. On the other hand, according to equation (3.3), we have:

$$p(x) = \sum_{i=0}^{n} f_i L_i(x) = \sum_{i=0}^{n} L_i(x)$$

By contradiction, suppose that $p(x) \equiv 1$ is not hold, then equation $p(x) - 1 = 0$ has the $n+1$ roots of $x_0, \ldots, x_n$, while according to the above discussion, $p \in \Pi_n$, so $p(x) - 1 = 0$ is a polynomial of at most degree n and cannot have $n+1$ roots.

As a result, the assumed statement is false and $p(x) \equiv 1$ and $\sum_{i=0}^{n} L_i(x) = 1$.

3.2.2 Problem

If $w(x) = \prod_{j=0}^{n} (x - x_j)$, show

$$L_i(x) = \frac{w(x)}{(x - x_i) w'(x_i)}$$

Solution: According to the definition of the derivative of the function in the point x_i, we have:

$$w'(x_i) = \lim_{x \to x_i} \frac{w(x) - w(x_i)}{x - x_i} = \lim_{x \to x_i} \frac{w(x)}{x - x_i} = \prod_{j=0, j \neq i}^{n} (x_i - x_j)$$

According to equation (3.4):

$$L_i(x) = \prod_{j=0, j \neq i}^{n} \left(\frac{x - x_j}{x_i - x_j} \right)$$

So, we can write:

$$L_i(x) = \frac{\displaystyle\prod_{j=0, j \neq i}^{n} (x - x_j)}{\displaystyle\prod_{j=0, j \neq i}^{n} (x_i - x_j)} = \frac{\displaystyle\prod_{j=0}^{n} (x - x_j)}{(x - x_i) \displaystyle\prod_{j=0, j \neq i}^{n} (x_i - x_j)} = \frac{w(x)}{(x - x_i) w'(x_i)}$$

3.2.3 Problem

Suppose $h(x) = L_i(x) + L_{i+1}(x)$ for $0 \le i < n$. Prove that for every x of the interval $[x_i, x_{i+1}]$ we have:

$$h(x) = L_i(x) + L_{i+1}(x) \ge 1$$

Solution: By contradiction, assume that:

$$\exists x \big(x \in (x_i, x_{i+1}) \Rightarrow L_i(x) + L_{i+1}(x) < 1, \, 0 \le i \le n-1 \big)$$

Consider the following partition:

$$x_0 < x_1 < \cdots < x_{i-1} < x_i < x < x_{i+1} < \cdots < x_{n-1} < x_n$$

According to the Rolle's theorem, for each of the subintervals of (x_i, x_{i+1}), $i = 0, 1, \ldots, i-1, i+2, \ldots, n-1$, $h'(x) = 0$ has one root, and since the number of these subintervals is $n-3$, then $h'(x) = 0$ has $n-3$ roots in the mentioned subintervals. Now according to the average value theorem, it can be written as follows:

$$\exists \beta_1 \left(\beta_1 \in (x_{i-1}, x_i) \,\&\, h'(\beta_1) = \frac{h(x_i) - h(x_{i-1})}{x_i - x_{i-1}} > 0 \right)$$

$$\exists \beta_2 \left(\beta_2 \in (x_i, x) \,\&\, h'(\beta_2) = \frac{h(x) - h(x_i)}{x - x_i} < 0 \right)$$

$$\exists \beta_3 \left(\beta_3 \in (x, x_{i+1}) \,\&\, h'(\beta_3) = \frac{h(x_{i+1}) - h(x)}{x_{i+1} - x} > 0 \right)$$

$$\exists \beta_4 \left(\beta_4 \in (x_{i+1}, x_{i+2}) \,\&\, h'(\beta_4) = \frac{h(x_{i+2}) - h(x_{i+1})}{x_{i+2} - x_{i+1}} < 0 \right)$$

According to the median value theorem, it can be written as follows:

$$\exists \theta_1 \big(\theta_1 \in (\beta_1, \beta_2) \,\&\, h'(\theta_1) = 0 \big)$$

$$\exists \theta_2 \big(\theta_2 \in (\beta_2, \beta_3) \,\&\, h'(\theta_2) = 0 \big)$$

$$\exists \theta_3 \big(\theta_3 \in (\beta_3, \beta_4) \,\&\, h'(\theta_3) = 0 \big)$$

Therefore, $h'(x) = 0$ has three other roots. As a result, totally, it will have n roots. But since $h(x)$ is of at most degree n, then $h'(x)$ is of at most degree $n-1$ and cannot have n roots. Therefore, the assumption is invalid and $h(x) \geq 1$.

3.2.4 Problem

Prove that Lagrange polynomials are linearly independent.
   Solution: Suppose that for every x:

$$a_0 L_0(x_j) + a_1 L_1(x_j) + \cdots + a_j L_j(x_j) + \cdots + a_n L_n(x_j) = 0$$

We should prove $a_0 = a_1 = \cdots = a_n = 0$.
   By substituting x_j in the above relation, we have:

$$a_0 L_0(x) + a_1 L_1(x) + \cdots + a_n L_n(x) = 0$$

Since the Lagrange polynomials satisfy the Kronecker delta condition, then the left hand of the above equation is equal to a_j. As a result:

$$a_j = 0, \quad j = 0, \ldots, n$$

3.2.5 Problem

Prove that the following set is linearly independent.
   Solution: Suppose that for every x,

$$\{1, (x - x_0), (x - x_0)(x - x_1), \ldots, (x - x_0) \ldots (x - x_{n-1})\}$$

We prove that $a_0 = a_1 = \cdots = a_n = 0$. By contradiction, we suppose that i is the first index for which $a_i \neq 0$. So, for every x, we have:

$$a_0 + a_1(x - x_0) + \cdots + a_i(x - x_0) \ldots (x - x_{i-1}) + \cdots + a_n(x - x_0) \ldots (x - x_{n-1}) = 0$$

Now if we put substitute x by x_i, we get:

$$a_i(x_i - x_0) \ldots (x_i - x_{i-1}) = 0$$

Because the points are mutually distinctive, then $a_{i=0}$ which is contrary to the assumption. As a result, the assumed statement is invalid and $a_0 = a_1 = \cdots = a_n = 0$.

3.3 Iterative Interpolation

These types of interpolations are also a special case of type (1). Suppose that there is a sequence of points $\left\{\left(x_{i_j}, f_{i_j}\right)\right\}_{j=0}^{n}$. It should be noted that the reason for the presence of two indices is that the order of the points is not important.

A polynomial $P_{i_0,\ldots,i_n}(x)$ of at most degree n is called a Neville interpolator for the above points, where

$$P_{i_0,\ldots,i_n}(x) = \frac{\left(x - x_{i_0}\right)P_{i_1,\ldots,i_n}(x) - \left(x - x_{i_n}\right)P_{i_0,\ldots,i_{n-1}}(x)}{x_{i_n} - x_{i_0}}$$

It is clear that Neville's method works with the two first and last indices of the data and is also invariant under permutation of indices. This method moves symmetrically and in the form of an isosceles triangle in the data table.

A polynomial $P_{i_0,\ldots,i_k,i_l,i_t}(x)$ of at most degree n is called **Aitken interpolator** in which

$$P_{i_0,\ldots,i_k,i_l,i_t}(x) = \frac{\left(x - x_{i_l}\right)P_{i_0,\ldots,i_k,i_t}(x) - \left(x - x_{i_t}\right)P_{i_0,\ldots,i_k,i_l}(x)}{x_{i_t} - x_{i_l}}$$

The Aitken method works with the last two indices of the data and moves in the data table as a right triangle.

By comparing these two methods in terms of error propagation, it can be claimed that the Aitken method has a more relative delete error because if x_i are approximate; at the denominator of the Aitken method, we will have the difference of approximate numbers with the same sign, and this increases the error propagation.

Because the use of computer programming is better for recursive relationships, it will be easier to work with Neville and Aitken methods. Also, in these two methods, in addition to the interpolation polynomial, the value of the function at the desired point is also obtained using the data table, and unlike the Lagrange method, the degree of the interpolation polynomial can be determined with simpler calculations.

3.3.1 Problem

Show that the Neville interpolator is invariant under the permutations of the indices.

Solution: Suppose that $i_0,\ldots,i_k$ and $j_0,\ldots,j_k$ are two permutations of indices. We prove:

$$P_{i_0,\ldots,i_k}(x) = P_{j_0,\ldots,j_k}(x)$$

by induction on k. Suppose $k = 1$, then:

$$P_{i_0, i_1}(x) = \frac{(x - x_{i_1})P_{i_0}(x) - (x - x_{i_0})P_{i_1}(x)}{x_{i_0} - x_{i_1}}$$

$$P_{j_0, j_1}(x) = \frac{(x - x_{j_1})P_{j_0}(x) - (x - x_{j_0})P_{j_1}(x)}{x_{j_0} - x_{j_1}}$$

Obviously, in the presence of only two points, $P_{j_0, j_1}(x) = P_{i_0, i_1}(x)$, so the initial case is true. Now we suppose that for any $k > 2$, the relation

$$P_{i_0, \ldots, i_k}(x) = \frac{(x - x_{i_k})P_{i_0, \ldots, i_{k-1}}(x) - (x - x_{i_0})P_{i_1, \ldots, i_k}(x)}{x_{i_0} - x_{i_k}}$$

is invariant under permutation of indices. We prove that

$$P_{i_0, \ldots, i_k, i_{k+1}}(x) = \frac{(x - x_{i_{k+1}})P_{i_0, \ldots, i_k}(x) - (x - x_{i_0})P_{i_1, \ldots, i_{k+1}}(x)}{x_{i_0} - x_{i_{k+1}}}$$

is invariant under permutation of indices. Given that we have $k + 2$ indices, then $(k + 2)!$ permutation can be written, that is, the number of states is equal to the permutation of a $(k + 1)$-member set with a single-member set. The permutation of the $(k + 1)$-member set, according to the induction assumption, is invariant under permutation of indices. The permutations of the $(k + 1)$-member set with the single-member set are also invariant under permutations of the indices according to the initial case.

3.4 Interpolation by Newton's Divided Differences

This type of interpolation is type (1) interpolation based on the Neville recursive method. The polynomial

$$p(x) = f_0 + (x - x_0)f[x_0, x_1] + \cdots + (x - x_0)\ldots(x - x_{n-1})f[x_0, \ldots, x_n]$$

is the Newton's interpolation polynomial where

$$f[x_0, \ldots, x_n] = \frac{f[x_1, \ldots, x_n] - f[x_0, \ldots, x_{n-1}]}{x_n - x_0}$$

are nth order divided differences between points $x_0, \ldots, x_n$.

It should be noted that the Newton's divided differences are invariant under permutations of the indices, and the previous calculations can still be used by adding a point. So, the coefficients of this interpolation polynomial are stable.

3.4.1 Problem

Suppose that $f(x)$ is a polynomial of degree n, then:

$$\forall k(k > n \Rightarrow f[x_0, \ldots, x_k] = 0)$$

Solution: We assume that $f(x) = x^n$ without losing generality. It is clear that $f(x)$ passes through interpolation points, so it can be considered as an interpolation polynomial. On the other hand, if we want to write the divided differences interpolation polynomial formula, due to the uniqueness of the interpolation polynomial, we have:

$$x^n = f_0 + (x - x_0) f[x_0, x_1] + \cdots + (x - x_0) \ldots (x - x_{n-1}) f[x_0, \ldots, x_n]$$
$$+ \cdots + (x - x_0) \ldots (x - x_{k-1}) f[x_0, \ldots, x_k]$$

Therefore, the statement is hold.

3.4.2 Problem

Prove that if $x_i = x_0 + ih$ for $i = 0, \ldots, n$, then:

$$\Delta^n f(x_0) = n! h^n f[x_0, \ldots, x_n]$$

Solution: We prove the above relation by induction on n.
   Initial case: Suppose $n = 1$.

$$\Delta f(x_0) = f(x_1) - f(x_0) = (x_1 - x_0) \frac{f(x_1) - f(x_0)}{x_1 - x_0} = hf[x_0, x_1]$$

Induction hypothesis: We suppose that the above relation holds for $n \geq 1$.
Induction step: We prove that it holds for $n + 1$ as well.

$$\Delta^{n+1} f(x_0) = \Delta^n f(x_1) - \Delta^n f(x_0)$$

$$= n! h^n f[x_1, \ldots, x_{n+1}] - n! h^n f[x_0, \ldots, x_n]$$

$$= n! h^n f(x_{n+1} - x_0) \frac{f[x_1, \ldots, x_{n+1}] - f[x_0, \ldots, x_n]}{x_{n+1} - x_0}$$

$$= (n+1)! h^{n+1} f[x_0, \ldots, x_{n+1}]$$

3.4.3 Problem

Prove that:

$$f[x_0,\ldots,x_n] = \sum_{i=0}^{n} \frac{f(x_i)}{\displaystyle\prod_{j=0, j\neq i}^{n} (x_i - x_j)}$$

Solution: According to the Lagrange interpolation formula, we have:

$$p(x) = \sum_{i=0}^{n} f_i L_i(x)$$

$$= \sum_{i=0}^{n} f_i \prod_{j=0, j\neq i}^{n} \frac{x - x_j}{x_i - x_j}$$

$$= \sum_{i=0}^{n} \prod_{j=0, j\neq i}^{n} (x - x_j) \frac{f_i}{\displaystyle\prod_{j=0, j\neq i}^{n} (x_i - x_j)}$$

On the other hand, according to Newton's interpolation formula, we have:

$$p(x) = f(x_0) + (x - x_0) f[x_0, x_1] + \cdots + (x - x_0)\ldots(x - x_{n-1}) f[x_0,\ldots,x_n]$$

Because the interpolation polynomial is unique, the coefficients of the leading terms must be equal, then:

$$f[x_0,\ldots,x_n] = \sum_{i=0}^{n} \frac{f(x_i)}{\displaystyle\prod_{j=0, j\neq i}^{n} (x_i - x_j)}$$

3.4.4 Problem

Suppose that $x_j = x_0 + jh$ for $j = 0,\ldots,n$. In this case, prove that:

$$\Delta^n f(x_0) = \sum_{i=0}^{n} (-1)^{n-i} C_n^i f(x_i)$$

Solution: Given that $x_j = x_0 + jh$, $\quad j = 0,\ldots,n$, then:

$$\prod_{j=0,j\neq i}^{n}(x_i - x_j) = \prod_{j=0,j\neq i}^{n}(i-j)h$$

$$= \prod_{j=0}^{i-1}(i-j)h \cdot \prod_{l=i+1}^{n}(i-l)h$$

$$= (h \cdot 2h \ldots ih)\big((-h)\cdot(-2h)\ldots(-(n-i)h)\big)$$

$$= (-1)^{n-1} i!(n-i)! h^n$$

According to the problem (3.4.3), we have:

$$f[x_0,\ldots,x_n] = \sum_{i=0}^{n}\frac{f(x_i)}{\prod\limits_{j=0,j\neq i}^{n}(x_i - x_j)}$$

$$= \sum_{i=0}^{n}\frac{f(x_i)}{(-1)^{n-i} i!(n-i)! h^n}$$

$$= \frac{1}{n!h^n}\sum_{i=0}^{n}(-1)^{n-i} C_n^i f(x_i)$$

So, according to the problem (3.4.2), we will have:

$$\Delta^n f(x_0) = n! h^n f[x_0,\ldots,x_n]$$

$$= \sum_{i=0}^{n}(-1)^{n-i} C_n^i f(x_i)$$

3.4.5 Problem

Assume that the derivatives of the function $f(x)$ in $I[x_0,\ldots,x_n]$ are continuously present up to n order. If $x_0,\ldots,x_n$ are distinct points, then:

$$f[x_0,\ldots,x_n] = \int_0^1 dt_1 \int_0^{t_1} dt_2 \cdots \int_0^{t_{n-1}} dt_n \times f^{(n)}\big(t_n(x_n - x_{n-1}) + \cdots + t_1(x_1 - x_0) + x_0\big)$$

where are $n \geq 1$ and $t_0 = 1$.

Solution: By induction on n, we prove the above relation.

Initial case: If $n = 1$, we must prove:

$$f[x_0,x_1] = \int_0^1 dt_1 \times f'\big(t_1(x_1 - x_0) + x_0\big)$$

For this purpose, we introduce a new integration variable named ξ for which

$$\xi = t_1(x_1 - x_0) + x_0, \quad dt_1 = \frac{d\xi}{x_1 - x_0}$$

The integration bounds change as follows:

$$t_1 = 0 \quad \Rightarrow \quad \xi = x_0$$
$$t_1 = 1 \quad \Rightarrow \quad \xi = x_1$$

So, we will have

$$\int_0^1 dt_1 \times f'\big(t_1(x_1 - x_0) + x_0\big) = \frac{1}{x_1 - x_0} \int_{x_0}^{x_1} d\xi \times f'(\xi)$$

$$= \frac{f(x_1) - f(x_0)}{x_1 - x_0}$$

$$= f[x_0, x_1]$$

Inductive hypothesis: Suppose that the relation holds for $n-1$, that is,

$$f[x_0, \ldots, x_{n-1}] = \int_0^1 dt_1 \int_0^{t_1} dt_2 \cdots \int_0^{t_{n-2}} dt_{n-1} \times f^{(n-1)}\big(t_{n-1}(x_{n-1} - x_{n-2}) + \cdots + t_1(x_1 - x_0) + x_0\big)$$

Induction step: We prove that the mentioned relation is hold for n.

$$\xi = t_n(x_n - x_{n-1}) + \cdots + t_1(x_1 - x_0) + x_0, \quad dt_n = \frac{d\xi}{x_n - x_{n-1}}$$

Therefore, the integration bounds will be as follows:

$$t_n = 0 \Rightarrow \xi = \xi_0 \equiv t_{n-1}(x_{n-1} - x_{n-2}) + \cdots + t_1(x_1 - x_0) + x_0$$
$$t_n = t_{n-1} \Rightarrow \xi = \xi_1 \equiv t_{n-1}(x_n - x_{n-2}) + t_{n-2}(x_{n-2} - x_{n-3}) + \cdots + t_1(x_1 - x_0) + x_0$$

In this case, the intermediate integral is as follows:

$$\int_0^{t_{n-1}} dt_n \times f^{(n)}\big(t_n(x_n - x_{n-1}) + \cdots + t_1(x_1 - x_0) + x_0\big)$$

$$= \int_{\xi_0}^{\xi_1} f^{(n)}(\xi) \frac{d\xi}{x_n - x_{n-1}} = \frac{f^{(n-1)}(\xi_1) - f^{(n-1)}(\xi_0)}{x_n - x_{n-1}}$$

we have:

$$\int_0^1 dt_1 \int_0^{t_1} dt_2 \cdots \int_0^{t_{n-2}} dt_{n-1} \times \frac{f^{(n-1)}(\xi_1) - f^{(n-1)}(\xi_0)}{x_n - x_{n-1}}$$

$$= \frac{f[x_0,\ldots,x_{n-2},x_n] - f[x_0,\ldots,x_{n-2},x_{n-1}]}{x_n - x_{n-1}}$$

$$= f[x_0,\ldots,x_n]$$

3.4.6 Point

1. If $f^{(n)}(x)$ is continuous on the interval $[a,b]$ and $y_0,\ldots,y_n$ are points in the $[a,b]$ and $x \in [a,b]$ is distinct from y_i, then

$$f[x,y_0,\ldots,y_n] = \frac{f[x,y_1,\ldots,y_n] - f[y_0,\ldots,y_n]}{x - y_0}$$

is a unique continuous generalization of the definition of divided differences.

2. If $\{x_i\}$ and $\{y_i\}$ are two sets of variables in $[a,b]$ where $x_i \neq y_j$, $0 \leq i \leq p, 0 \leq j \leq q$, $0 \leq p,q \leq m$ and $f^{(m)}(x)$ is continuous on the interval $[a,b]$, then:

$$f[x_0,\ldots,x_p,y_0,\ldots,y_q] = g[x_0,\ldots,x_p] = h[y_0,\ldots,y_q]$$

where

$$g(x) \equiv f[x,y_0,\ldots,y_q], \quad h(y) \equiv f[x_0,\ldots,x_p,y]$$

gives the unique continuous generalization of the definition of divided differences.

3. If $f^{(n)}(x)$ is continuous in $[a,b]$ and $x_0,\ldots,x_n$ are points in $[a,b]$, then:

$$f[x_0,\ldots,x_n] = \frac{f^{(n)}(\xi)}{n!}, \quad \xi \in I[x_0,\ldots,x_n]$$

3.4.7 Problem

If function $f(x)$ has continuous derivatives up to order m on the interval $[a,b]$ and $\{x_i\},\{y_j\}$, and $\{z_k\}$ are sets of variables in $[a,b]$, so that $x_i \neq y_j$, $x_i \neq z_k$, $y_j \neq z_k$, $0 \leq i \leq p, 0 \leq j \leq q, 0 \leq k \leq r$ and $0 \leq p,q,r \leq m$, then prove

$$f\left[x_0,\ldots,x_p,y_0,\ldots,y_q,z_0,\ldots,z_r\right] = \frac{1}{p!q!r!}\left[\frac{\partial^p}{\partial y^q}\cdot\frac{\partial^q}{\partial y^q}\cdot\frac{\partial^r}{\partial z^r}f\left[x,y,z\right]\right]_{(\xi,\eta,\gamma)}$$

where

$$\gamma \in I\left[z_0,\ldots,z_r\right],\quad \eta \in I\left[y_0,\ldots,y_q\right],\quad \xi \in I\left[x_0,\ldots,x_p\right]$$

Solution: Suppose that:

$$g(x) \equiv f\left[x,y_0,\ldots,y_q,z_0,\ldots,z_r\right]$$
$$h(y) \equiv f\left[x,y,z_0,\ldots,z_r\right]$$
$$k(z) \equiv f\left[x,y,z\right]$$

Using the points (2) and (3) and their appropriate generalization for the set of variables $\{x_i\}, \{y_j\}$ and $\{z_k\}$, we will have:

$$g\left[x_0,\ldots,x_p\right] = \frac{1}{p!}\left[\frac{\partial^p}{\partial x^p}g(x)\right]$$

$$g(x) = h\left[y_0,\ldots,y_q\right] = \frac{1}{q!}\left[\frac{\partial^q}{\partial y^q}h(y)\right]_{y=\eta}$$

$$h(x) = k\left[z_0,\ldots,z_r\right] = \frac{1}{r!}\left[\frac{\partial^r}{\partial y^r}k(z)\right]_{z=\gamma}$$

According to the above relations, the statement will be true.

3.4.8 Problem

Suppose that $x_0,\ldots,x_n$ are mutual distinct points. Prove that the coefficients $a_0,\ldots,a_n$ in interpolation polynomials $p \in P_n$ are continuously dependent on the values of $y_0,\ldots,y_n$.

   Solution: Suppose that

$$p(x) = a_0 + (x-x_0)a_1 + \cdots + (x-x_0)\cdots(x-x_{n-1})a_n$$

is the polynomial interpolation of the function f at the points $x_0,\ldots,x_n$. Therefore, given that we must have:

$$f(x_i) = p(x_i),\quad i = 0,\ldots,n$$

a_i s can be calculated as:

$$f(x_0) = p(x_0) = a_0$$

$$f(x_1) = p(x_1) = a_0 + a_1(x_1 - x_0)$$

$$= f(x_0) + a_1(x_1 - x_0)$$

So:

$$a_1 = \frac{f(x_1) - f(x_0)}{x_1 - x_0}$$

$$f(x_2) = p(x_2) = a_0 + a_1(x_2 - x_0) + a_2(x_2 - x_0)(x_2 - x_1)$$

$$= f(x_0) + \frac{f(x_1) - f(x_0)}{x_1 - x_0}(x_2 - x_0) + a_2(x_2 - x_0)(x_2 - x_1)$$

Then:

$$a_2 = \frac{f(x_2) - f(x_0)}{(x_2 - x_0)(x_2 - x_1)} - \frac{f(x_1) - f(x_0)}{(x_1 - x_0)(x_2 - x_1)}$$

Given that $y_i \equiv f(x_i)$, therefore

$$a_0 = y_0$$

$$a_1 = \frac{y_1 - y_0}{x_1 - x_0}$$

$$a_2 = \frac{y_2 - y_0}{(x_2 - x_0)(x_2 - x_1)} - \frac{y_1 - y_0}{(x_1 - x_0)(x_2 - x_1)}$$

Similarly, by continuing the same process, the other a_i are obtained in terms of the points and y_i.

3.4.9 Point

The operator Δ is a leading operator and is defined as follows:

$$\Delta f_i = f_{i+1} - f_i$$

3.4.10 Problem

Prove that

$$\Delta\big(f(x)\cdot g(x)\big) = f(x)\cdot\Delta g(x) + \Delta f(x)\cdot g(x+h)$$

Solution:

$$\Delta\big(f(x)\cdot g(x)\big) = f(x+h)\cdot g(x+h) - f(x)\cdot g(x)$$

$$= f(x+h)\cdot g(x+h) - f(x)\cdot g(x) + f(x)\cdot g(x+h) = f(x)\cdot g(x+h)$$

$$= g(x+h)\big(f(x+h) - f(x)\big) + f(x)\big(g(x+h) - g(x)\big)$$

$$= g(x+h)\cdot\Delta f(x) + f(x)\cdot\Delta g(x)$$

3.4.11 Problem

If $f(x) = \dfrac{1}{x+c}$, where c is a constant (real or complex) number, prove that:

$$f[x_0,\ldots,x_n] = \frac{(-1)^n}{(x_0+c)\ldots(x_n+c)}$$

Solution: We prove it by induction on n.
   Initial case: Suppose $n = 1$:

$$f[x_0,x_1] = \frac{f(x_1) - f(x_0)}{x_1 - x_0}$$

$$= \frac{\dfrac{1}{x_1+c} - \dfrac{1}{x_0+c}}{x_1 - x_0}$$

$$= \frac{x_0 - x_1}{(x_0+c)(x_1+c)} \cdot \frac{1}{x_1 - x_0}$$

$$= \frac{-1}{(x_0+c)(x_1+c)}$$

Induction hypothesis: Suppose that the above relation is hold for $n = k$.
   Induction step: We prove that it holds for $n = 2$ as well.

According to the induction hypothesis, we know:

$$f[x_1,\ldots,x_{k+1}] = \frac{(-1)^k}{(x_1+c)\ldots(x_{k+1}+c)}$$

$$f[x_0,\ldots,x_k] = \frac{(-1)^k}{(x_0+c)\ldots(x_k+c)}$$

So,

$$f[x_0,\ldots,x_k,x_{k+1}] = \frac{f[x_1,\ldots,x_{k+1}] - f[x_0,\ldots,x_k]}{x_{k+1},x_0}$$

$$= \left(\frac{(-1)^k}{(x_1+c)\ldots(x_{k+1}+c)} = \frac{(-1)^k}{(x_0+c)\ldots(x_k+c)}\right) \times \left(\frac{1}{x_{k+1}-x_0}\right)$$

$$= \frac{(-1)^k(x_0-x_{k+1})}{(x_0+c)\ldots(x_{k+1}+c)} \cdot \frac{1}{x_{k+1}-x_0}$$

$$= \frac{(-1)^{k+1}}{(x_0+c)\ldots(x_{k+1}+c)}$$

Therefore, the statement is true.

3.4.12 Problem

Suppose that $x_0,\ldots,x_n$ are distinct points, and $f \in c^n(I(x_0,\ldots,x_n))$, in this case:

$$f[x_0,\ldots,x_n] = \int\cdots\int_{\tau n} f^{(n)}(t_0 x_0 + \cdots + t_n x_n)dt_1\,dt_2\ldots dt_n$$

where are $t_0 = 1 - \sum_{j=1}^{n} t_j$ and $\tau_n = \left\{(t_0,\ldots,t_n)\Big| t_j \geq 0, \sum_{j=1}^{n} t_j \leq 1\right\}$.

Solution: We prove the statement by induction on n.
Initial case: Suppose $n = 1$ with $\tau_1 = [0,1]$ and $t_0 = 1 - t_1$, then

$$\int_{\tau_1} f'(t_0 x_0 + t_1 x_1)dt_1 = \int_0^1 f'(x_0 + t_1(x_1 - x_0))dt_1$$

$$= \frac{f(x_1) - f(x_0)}{x_1 - x_0}$$

$$= f[x_0,x_1]$$

Induction hypothesis: Assume that the statement holds for $n = k - 1$.

Induction step: We prove that the sentence holds for $n = k$ as well: Note that $t_0 = 1 - t_1 - \cdots - t_k$ and $\tilde{t}_k = 1 - t_1 - \cdots - t_{k-1}$

So,

$$\int \cdots \int_{\tau_k} f^{(k)}(t_0 x_0 + \cdots + t_k x_k) dt_1 \ldots dt_k$$

$$= \int \cdots \int_{\tau_{k-1}} \int_0^{\tilde{t}_k} f^{(k)}(x_0 + t_1(x_1 - x_0) + \cdots + t_k(x_k - x_0)) dt_k \, dt_1 \ldots dt_{k-1}$$

$$\int \cdots \int_{\tau_{k-1}} \frac{1}{x_k - x_0} [f^{(k-1)}(x_0 + t_1(x_1 - x_0) + \cdots + \tilde{t}_k(x_k - x_0))$$

$$- f^{(k-1)}(\tilde{t}_k x_0 + t_1 x_1 + \cdots + t_{k-1} x_{k-1}) dt_1 \ldots dt_{k-1}$$

$$= \int_0^1 \int_0^{1-t_1} \cdots \int_0^{1-t_1-\cdots-t_{k-2}} \frac{f^{(k-1)}(t_1 x_1 + \cdots + \tilde{t}_k x_k)}{x_k - x_0} dt_1 \ldots dt_{k-1}$$

$$- \int \cdots \int \frac{f^{(k-1)}(\tilde{t}_k x_0 + t_1 x_1 + \cdots + t_{k-1} x_{k-1})}{x_k - x_0} dt_1 \ldots dt_{k-1}$$

$$= \frac{f[x_1, \ldots, x_k] - f[x_0, \ldots, x_{k-1}]}{x_k - x_0}$$

$$= f[x_0, \ldots, x_k]$$

3.4.13 Problem

Suppose $x_0 = x_1 = \cdots = x_n$. Show using the assumptions of problem (3.3.12):

$$\lim_{x_j \to x_0} f[x_0, \ldots, x_n] = \frac{f^{(n)}(x_0)}{n!}$$

Solution:

$$\lim_{x_j \to x_0} f[x_0, \ldots, x_n] = \int \cdots \int_{\tau_n} f^{(n)}((t_0 + \cdots + t_n) x_0) dt_1 \ldots dt_n$$

$$= \int \cdots \int_{\tau_n} f^{(n)}(x_0) dt_1 \ldots dt_n$$

$$= f^{(n)}(x_0) \int \cdots \int_{\tau_n} dt_1 \ldots dt_n$$

$$= \frac{f^{(n)}(x_0)}{n!}$$

3.4.14 Problem

Show that if α is constant and positive, and $f(x) = \alpha^x$, then in what conditions

$$\Delta f_i = f_i$$

always holds.

Solution: We have to find the condition for $\Delta f_i = f_{i+1} - f_i = f_i$.

$$f_{i+1} = 2f_i$$
$$\Rightarrow \alpha^{x_i+1} = 2\alpha^{x_i}$$
$$\Rightarrow \ln \alpha^{x_i+1} = \ln 2\alpha^{x_i}$$
$$\Rightarrow x_{i+1} \ln \alpha = \ln 2 + x_i \ln \alpha$$
$$\Rightarrow (x_{i+1} - x_i) \ln \alpha = \ln 2$$
$$\Rightarrow x_{i+1} = x_i + \frac{\ln 2}{\ln \alpha}$$

So, by choosing $h = \dfrac{\ln 2}{\ln \alpha}$, $\Delta f_i = f_i$ will always hold.

3.5 Exercise

1. Explain that, if components $f(x_n), \ldots, f(x_0)$ be the only information about the function $f(x)$, then the error of $R_n(\bar{x}) = f(\bar{x}) - p(\bar{x})$ could not be justified.

2. Find coefficients a_0, a_1, a_2, a_3 from the polynomial

$$p(x) = a_0 + a_1(x - y) + a_2(x - y)^2 + a_3(x - y)^3$$

if

$$p(y) = f_y, \quad p'(y) = f_y'$$

$$p(z) = f_z, \quad p'(z) = f_z'$$

and $y \neq z$.

3. If $p(x)$ be the same polynomial as in (2), find $p\left(\dfrac{y+z}{2}\right)$ given $y, z, f_y, f_z, f_y', f_z'$.

4. Show that the formulas of Lagrange interpolation conclude directly from Newton's divided difference.

5. We define

$$\frac{d}{dt} f[t_0,\ldots,t_k,t] = f[t_0,\ldots,t_k,t,t]$$

prove that if $u_n,\ldots,u_1$ are differentiable functions, then

$$\frac{d}{dt} f[t_0,\ldots,t_k,u_1,\ldots,u_n] = \sum_{j=1}^{n} f[t_0,\ldots,t_k,u_1,\ldots,u_n,u_j]\frac{du_j}{dt}$$

4

Advanced Interpolation

4.1 Hermit Interpolation

In this section, we introduce an interpolation that operates in a multicriteria manner. In this interpolator, we assume that at the interpolation points, the value of the function f is equal to the value of the interpolation function, as well as the value of the finite derivatives of f is equal to the value of the finite derivatives of the interpolation function, and this is the reason for smoothness and accuracy of the interpolation function.

Suppose that x_i s are real numbers for $i = 0,\dots,n$, so that:

$$x_0 \le x_i \le \cdots \le x_n$$

Therefore, for $i = 0,\dots,m$, there are points like ξ_i with a new arrangement so that:

$$\xi_0 < \xi_1 < \cdots < \xi_m$$

So, we consider the interpolation points as follows:

$$\left(\xi_i, y_i^{(k)}\right), \quad i = 0,\dots,m, \quad k = 0,\dots,n_i - 1, \quad n_i \in \mathbb{N}$$

The polynomial of at most degree n of

$$p(x) = \sum_{i=0}^{m} \sum_{k=0}^{n_i-1} y_i^{(k)} L_{ik}(x) \tag{4.1}$$

is called Hermit interpolation polynomial if it satisfies the following interpolation conditions:

$$p^{(k)}(\xi_i) = y_i^{(k)}, \quad i = 0,\dots,m, \quad k = 0,\dots,n_i - 1$$

DOI: 10.1201/9781003218173-4

Note that $\sum_{i=0}^{m} n_i = n+1$, and if for every i, $n_i = 1$, then we have the same inter-

polation problems with $m+1$ distinct point of (ξ_i, y_i).

In the equation (4.1), for $i = 0,\ldots,m$ and $k = 0,\ldots,n_i - 1$, L_{ik}s are defined as follows:

$$L_{ik}(x) = l_{ik}(x) - \sum_{v=k+1}^{n_i-1} l_{ik}^{(v)}(\xi_i) L_{iv}(x) \tag{4.2}$$

where

$$l_{ik}(x) = \frac{(x-\xi_i)^k}{k!} \prod_{j=0, j\neq i}^{m} \left(\frac{x-\xi_i}{\xi_i - \xi_i} \right)^{n_j} \tag{4.3}$$

In a special case, if for every i, $n_i = 2$, the interpolator is called simple Hermit, where

$$f(\xi_i) = p(\xi_i), \quad f'(\xi_i) = p'(\xi_i), \quad i = 0,\ldots,m$$

And so, it can be said that simple Hermit interpolation polynomials are obtained as follows:

$$p(x) = \sum_{i=0}^{m} y_i H_i(x) + \sum_{i=0}^{m} y_i' Q_i(x)$$

where

$$H_i(x) = \left(1 - 2(x - \xi_i) L_i'(\xi_i) \right) L_i^2(x)$$

$$Q_i(x) = (x - \xi_i) L_i^2(x)$$

In Hermit interpolation, due to the presence of consecutive derivatives, the slope of the function is equal to the slope of the interpolator function at the interpolation points, and the presence of derivatives causes the number of points to be greater, so the degree of interpolation polynomials is higher. Therefore, the interpolator is more accurate and converges to the function faster. It should also be noted that if the derivatives are not present sequentially at one point, this interpolator cannot be used.

4.1.1 Problem

Show that in the Hermite interpolation, $L_{i,k}$ s form an independent linear set.

Solution: We should prove that for any arbitrary combination of $L_{i,k}(x)$ s that is set to zero, the coefficients are always equal to zero. So, we assume that:

$$\sum_{i=0}^{m}\sum_{k=0}^{n_i-1} f_i^{(k)} L_{i,k}(x) = 0$$

is an arbitrary combination of $L_{i,k}$ s that is equal to zero, so we should prove that

$$\forall i \; \forall k \left(i = 0,\ldots,m, \quad k = 0,\ldots,n_i - 1 \Rightarrow f_i^{(k)} = 0 \right)$$

As $p(x)$ is equal to zero for any x, so it will be zero for $x_0,\ldots,x_n$ and also for $\xi_0,\ldots,\xi_m$. Therefore, the consecutive derivatives of $p(x)$ at these points are also equal to zero:

$$p^{(\sigma)}(\xi_j) = \sum_{i=0}^{m}\sum_{k=0}^{n_i-1} f_i^{(k)} L_{i,k}^{(\sigma)}(\xi_i) = 0, \quad j = 0,\ldots,m, \quad \sigma = 0,\ldots,n_j - 1$$

If we expand the first summation of the above relation, we will have:

$$\sum_{k=0}^{n_0-1} f_0^{(k)} L_{0,k}^{(\sigma)}(\xi_j) + \cdots + \sum_{k=0}^{n_j-1} f_j^{(k)} L_{j,k}^{(\sigma)}(\xi_j) + \cdots + \sum_{k=0}^{n_m-1} f_m^{(k)} L_{m,k}^{(\sigma)}(\xi_j) = 0$$

As $L_{i,k}^{(\sigma)}(\xi_i)$ satisfies the Kronecker Delta, then in the above relation, all $L_{i,k}^{(\sigma)}(\xi_i)$ s are equal to zero except $L_{j,k}^{(\sigma)}(\xi_j)$. So, we have:

$$\sum_{k=0}^{n_j-1} f_j^{(k)} L_{j,k}^{(\sigma)}(\xi_j) = 0$$

Now if we expand the summation, we will have:

$$f_j^{(0)} L_{j,0}^{(\sigma)}(\xi_j) + \cdots + f_j^{(\sigma)} L_{j,\sigma}^{(\sigma)}(\xi_j) + \cdots + f_j^{(n_j-1)} L_{j,n_j-1}^{(\sigma)}(\xi_j) = 0$$

Again, according to the Kronecker Delta condition, all $L_{j,k}^{(\sigma)}(\xi_j)$ s are equal to zero and $L_{j,\sigma}^{(\sigma)}(\xi_j) = 1$.
So,

$$f_j^{(\sigma)} = 0, \quad j = 0,\ldots,m \quad \sigma = 0,\ldots,n_j - 1$$

4.1.2 Problem

Show that in the Hermite interpolation, $L_{i,k}(x)$ s is of at most degree n.
   Solution: According to the equations (4.2) and (4.3):

$$L_{i,k}(x) = l_{i,k}(x) - \sum_{v=k+1}^{n_j-1} l_{i,k}^{(v)}(\xi_i) L_{i,v}(x)$$

and

$$l_{i,k}(x) = \frac{(x-\xi_i)^k}{k!} \prod_{j=0, j\neq i}^{m} \left(\frac{x-\xi_j}{\xi_i-\xi_j}\right)^{n_j}, \quad k = 0,\ldots,n_i - 1$$

According to the above definitions, we have:

$$\deg(l_{i,k}) = n+1-n_i+k \leq n+1-n_i+n_i-1 = n$$

Therefore, the degree of $L_{i,k}$, regardless of any degree achieved for $L_{i,v}(x)$ is at most n.

4.1.3 Problem

For points $\{(x_0, f_i)\}_{i=0}^{n}$, prove that the n-th order derivative of the interpolating function at x_0 is equal to the product of the interpolator coefficients and n_i.
   Solution: We know that

$$f(x) = f_0 + (x-x_0)f_{01} + \cdots + (x-x_0)\ldots(x-x_{n-1})f_{01\ldots n}$$
$$+ (x-x_0)\ldots(x-x_n)f[x,x_0,\ldots,x_n]$$
$$= p(x) + (x-x_0)\ldots(x-x_n)f[x,x_0,\ldots,x_n]$$

where

$$f_{01\ldots n} := f[x_0,\ldots,x_n]$$

Therefore, it can be said that:

$$(x-x_0)\ldots(x-x_n)\frac{f^{(n+1)}(\xi_x)}{(n+1)!} = (x-x_0)\ldots(x-x_n)f[x,x_0,\ldots,x_n]$$

where $\xi_x \in I[x_0,\ldots,x_n]$. As a result:

$$\frac{f^{(n+1)}(\xi_x)}{(n+1)!} = f[x, x_0, \ldots, x_n]$$

The latter relation holds for every n and every set of $x, x_0, \ldots, x_n$ numbers, provided that the $(n+1)$-th derivative of f exists. By removing point x_0, we will have:

$$f[x, x_1, \ldots, x_n] = \frac{f^{(n)}(\eta_x)}{n!}, \quad \eta_x \in I[x, x_1, \ldots, x_n]$$

If we substitute x by x_0, then:

$$\lim_{x \to x_0} f[x, x_1, \ldots, x_n] = f[x_0, x_1, \ldots, x_n]$$

On the other hand

$$\forall x (x \to x_0 \Rightarrow \eta_x \to x_0)$$

So,

$$\lim_{x \to x_0} \frac{f^{(n)}(\eta_x)}{n!} = \frac{f^{(n)}(x_0)}{n!}$$

and it can be concluded that

$$f^{(n)}(x_0) = f[x_0, x_1, \ldots, x_n] \cdot n!$$

Because the lengths of all points are assumed to be equal, a Hermite interpolator must be used to interpolate them. According to the definition of the Hermite interpolating function, we have:

$$p^{(k)}(\xi_i) = f_i^{(k)}, \quad i = 0, 1, \ldots, m, \quad k = 0, 1, \ldots, n_i - 1$$

According to above, $m = 0$, $\xi_0 = x_0$ and $n_0 = n+1$, therefore

$$p^{(n)}(x_0) = f_0^{(n)}$$

As a result

$$p^{(n)}(x_0) = f[x_0, x_1, \ldots, x_n] \cdot n!$$

4.1.4 Problem

Suppose that $p_n(x) = \alpha_0 + \alpha_1 x + \cdots + \alpha_n x^n$, $\alpha_n \neq 0$ are polynomials of degree n and $x_0, \ldots, x_k$ are $k+1$ distinct points. Prove that:

$$p_n[x_0, \ldots, x_k] = \begin{cases} \alpha_n & k = n \\ 0 & k > n \end{cases}$$

Solution: Given that

$$\frac{d^n p_n(x)}{dx^n} = n! \alpha_n$$

Also, according to

$$f[x_0, \ldots, x_k] = \frac{f^{(k)}(\xi)}{k!}, \quad \xi \in I[x_0, \ldots, x_k]$$

It can be concluded from the problem (4.1.3) that:
  If $k = n$, then

$$p_n[x_0, \ldots, x_k] = \alpha_n$$

and if $k > n$, as for the degree of p_n, the derivative becomes zero, then:

$$p_n[x_0, \ldots, x_k] = 0$$

So,

$$p_n[x_0, \ldots, x_k] = \begin{cases} \alpha_n & k = n \\ 0 & k > n \end{cases}$$

4.2 Fractional Interpolation

In this section, we will investigate on how to interpolate functions if they have poles. Obviously, it is not possible to interpolate these functions with the methods mentioned in the previous sections, so if it is a fractional function, its best interpolators are asymptotical interpolating functions.

Now we want to introduce the general form of an asymptotical interpolator. Suppose that (x_i, f_i) s are interpolation points for $i = 0, 1, \ldots, \mu + v$. In this case, we define the fractional function $\phi^{\mu,v}$ as follows:

$$\varphi^{\mu,\upsilon}(x) = \frac{p^{\mu}(x)}{q^{\upsilon}(x)} = \frac{a_0 + a_1 x + \cdots + a_{\mu} x^{\mu}}{b_0 + b_1 x + \cdots + b_{\upsilon} x^{\upsilon}} \tag{4.4}$$

According to the above relation, it is clear that the unknown parameters of the fractional function of (4.4) are $a_0, \ldots, a_{\mu}, b_0, \ldots, b_{\upsilon}$, so that their number is $\mu + \upsilon + 2$. Therefore, it is sufficient to determine the unknown parameters in such a way that (4.4) become the interpolating function. In order for the fractional function (4.4) to interpolate the above-mentioned points, the following interpolation problem must be established:

$$\phi^{\mu,\upsilon}(x_i) = f_i, \quad i = 0, \ldots, \mu + \upsilon \tag{4.5}$$

If $\phi^{\mu,\upsilon}(x_i) \neq 0$, we can write:

$$\phi^{\mu,\upsilon}(x_i) = \frac{p^{\mu}(x_i)}{q^{\upsilon}(x_i)} = f_i, \quad i = 0, \ldots \mu + \upsilon$$

In this case,

$$p^{\mu}(x_i) - f_i q^{\upsilon}(x_i) = 0, \quad i = 0, \ldots, \mu + \upsilon \tag{4.6}$$

where the relation (4.6) is a homogenized system of (4.5). It is clear that the system (4.6) has $\mu + \upsilon + 2$ unknown and $\mu + \upsilon + 1$ equations, so to solve (4.6), we need a known parameter. For this purpose, we assume one of the denominator parameters as a known parameter and we obtain the fractional function (4.4) by solving the system (4.6).

4.2.1 Problem

Calculate the fractional interpolation error.

Solution: Suppose that $\{(x_i, f_i)\}_{i=1}^{\mu+\upsilon+1}$ are interpolation points where $x_i \in (a, b)$. We know that

$$f(x) = \frac{p(x)}{q(x)} + R(x)$$

Suppose $\alpha_1, \ldots, \alpha_t$ are inaccessible points in $[a, b]$ with the iteration order of $r_1, \ldots, r_t$, where $\displaystyle\sum_{i=1}^{t} r_i = m$ and $\deg(q(x)) \geq m$. We define:

$$\varphi(x) = (x - \alpha_1)^{r_1} \ldots (x - \alpha_t)^{r_t}$$

So, $f(x) \cdot \varphi(x)$ is defined on the interval $[a, b]$. We determine ψ such that $Q(x) = \psi(x) \cdot \varphi(x)$ and $\deg(Q(x)) = \deg(q(x))$. We study two cases.

a. $n = 2h \Rightarrow \mu + v + 1 = 2h \Rightarrow \mu + v = 2h - 1$

Therefore, in the fractional interpolation relation of $\dfrac{p(x)}{q(x)}$, p is of degree h and q is of degree $h - 1$. So, it can be said that

$$\deg(p(x) \cdot Q(x)) = 2h - 1 = n - 1$$

b. $n = 2h + 1 \Rightarrow \mu + v + 1 = 2h + 1 \Rightarrow \mu + v = 2h$

Therefore, in the fractional interpolation relation of $\dfrac{p(x)}{q(x)}$, p is of degree h and q is of degree h. So, we have

$$\deg(p(x) \cdot Q(x)) = 2h = n - 1$$

Now we define:

$$R(x) = K \cdot \frac{(x - x_1)...(x - x_n)}{Q(x) \cdot q(x)}$$

So,

$$f(x) = \frac{p(t)}{q(t)} + K \cdot \frac{(t - x_1)...(t - x_n)}{Q(t) \cdot q(t)}$$

We take the common denominator from above equation and put:

$$w(t) = f(t) \cdot Q(t) \cdot q(t) - p(t) \cdot Q(t) - K \cdot (t - x_1)...(t - x_n)$$

We determine K so that $x_1, ..., x_n$ and $\bar{x}$ are the roots of $w(t) = 0$, so according to the Rolle's theorem, $w'(t) = 0$ has n roots and... and finally $w^{(n)}(t) = 0$ has at least one root. If we call that root, η, we will have:

$$w^{(n)}(\eta) = 0$$

So,

$$w^{(n)}(\eta) = \left[\frac{d^n}{dt^n} \left(f(t) \cdot Q(t) \cdot q(t) \right) \right]_\eta - 0 - K \cdot n! = 0$$

Therefore,

$$K = \frac{\left[\dfrac{d^n}{dt^n} \left(f(t) \cdot Q(t) \cdot q(t) \right) \right]_\eta}{n!}$$

So, it can be concluded that:

$$R(x) = \frac{(x - x_1)...(x - x_n)}{Q(x) \cdot q(x) \cdot n!} \cdot \left[\frac{d^n}{dt^n} \left(Q(t) \cdot f(t) \cdot q(t) \right) \right]_\eta$$

4.2.2 Problem

Suppose the fractional function $\phi^{m,n}(x)$ is the answer of the following homogenized system for nodal points of $\{(x_k, f_k)\}_{k=0}^{m+n}$:

$$\left(a_0 + a_1 x_k + \cdots + a_m x_k^m \right) - f_k \left(b_0 + b_1 x_k + \cdots + b_n x_k^n \right) = 0$$

Show that $\phi^{m,n}(x)$ can be represented by the following determinant:

$$\phi^{m,n}(x) = \frac{\left| f_k \quad (x_k - x)...(x_k - x)^m (x_k - x) f_k ...(x_k - x)^n f_k \right|_{k=0}^{m+n}}{\left| 1 \quad (x_k - x)...(x_k - x)^m (x_k - x) f_k ...(x_k - x)^n f_k \right|_{k=0}^{m+n}}$$

In the above relation, the following representation is used for determinants:

$$\left| \alpha_k, ..., \xi_k \right|_{k=0}^{m+n} = \begin{vmatrix} \alpha_0 & \cdots & \xi_0 \\ \alpha_1 & \cdots & \xi_1 \\ \vdots & \vdots & \vdots \\ \alpha_{m+n} & \cdots & \xi_{m+n} \end{vmatrix}$$

Solution: we should prove that

$$\phi^{m,n}(x_i) = f_i, \quad i = 0, 1, ..., m+n$$

$$\phi^{m,n}(x_i) = \frac{\begin{vmatrix} f_0 & \theta_0 & \cdots & \theta_0^m & \theta_0 f_0 & \cdots & \theta_0^n f_0 \\ \vdots & \vdots & \vdots & \vdots & \vdots & \vdots & \vdots \\ f_i & \theta_i & \cdots & \theta_i^m & \theta_i f_i & \cdots & \theta_i^n f_i \\ \vdots & \vdots & \vdots & \vdots & \vdots & \vdots & \vdots \\ f_{m+n} & \theta_{m+n} & \cdots & \theta_{m+n}^m & \theta_{m+n} f_{m+n} & \cdots & \theta_{m+n}^n f_{m+n} \end{vmatrix}}{\begin{vmatrix} 1 & \theta_0 & \cdots & \theta_0^m & \theta_0 f_0 & \cdots & \theta_0^n f_0 \\ \vdots & \vdots & \vdots & \vdots & \vdots & \vdots & \vdots \\ 1 & \theta_i & \cdots & \theta_i^m & \theta_i f_i & \cdots & \theta_i^n f_i \\ \vdots & \vdots & \vdots & \vdots & \vdots & \vdots & \vdots \\ 1 & \theta_{m+n} & \cdots & \theta_{m+n}^m & \theta_{m+n} f_{m+n} & \cdots & \theta_{m+n}^n f_{m+n} \end{vmatrix}}$$

where $\theta_j = x_j - x$, $j = 0,1,\ldots,m+n$. By considering $x = x_i$, it can be seen that in ith row of the determinants of the denominator and numerator, all elements are equal to zero expect a_{i1} that is $a_{i1} = f_i$ for numerator and $a_{i1} = 1$ for denominator. Now if we expand the determinants on the first column and the second row, we will have:

$$\phi^{m,n}(x_i) = \frac{\begin{vmatrix} \theta_{0i} & \cdots & \theta_{0i}^m & \theta_{0i}f_0 & \cdots & \cdots & \theta_{0i}^n f_0 \\ \vdots & \vdots & \vdots & \vdots & \vdots & \vdots & \vdots \\ \theta_{i-1i} & \cdots & \theta_{i-1i}^m & \theta_{i-1i}f_{i-1} & \cdots & \cdots & \theta_{i-1i}^n f_{i-1} \\ \theta_{i+1i} & \cdots & \theta_{i+1i}^m & \theta_{i+1i}f_{i+1} & \cdots & \cdots & \theta^m_{i+1i}f_{i+1} \\ \cdots & \vdots & \vdots & \vdots & \vdots & \vdots & \vdots \\ \theta_{m+ni} & \cdots & \theta_{m+ni}^m & \theta_{m+ni}f_{m+n} & \cdots & \cdots & \theta_{m+ni}^n f_{m+n} \end{vmatrix}}{\begin{vmatrix} \theta_{0i} & \cdots & \theta_{0i}^m & \theta_{0i}f_0 & \cdots & \cdots & \theta_{0i}^n f_0 \\ \vdots & \vdots & \vdots & \vdots & \vdots & \vdots & \vdots \\ \theta_{i-1i} & \cdots & \theta_{i-1i}^m & \theta_{i-1i}f_{i-1} & \cdots & \cdots & \theta_{i-1i}^n f_{i-1} \\ \theta_{i+1i} & \cdots & \theta_{i+1i}^m & \theta_{i+1i}f_{i+1} & \cdots & \cdots & \theta^m_{i+1i}f_{i+1} \\ \cdots & \vdots & \vdots & \vdots & \vdots & \vdots & \vdots \\ \theta_{m+ni} & \cdots & \theta_{m+ni}^m & \theta_{m+ni}f_{m+n} & \cdots & \cdots & \theta_{m+ni}^n f_{m+n} \end{vmatrix}}$$

In the above relation, $\theta_{ji} = x_j - x_i$.

On the other hand, assuming that one of the denominator coefficients such as b_0 is non-zero, the following result is obtained by dividing both nominator and denominator by b_0:

$$\phi^{m,n}(x) = \frac{a_0' + a_1'x + \cdots + a_m'x^m}{1 + b_1'x + \cdots + b_n'x^n}$$

$$\phi^{m,n}(x_k) = f_k, \quad k = 0,\ldots,m+n$$

The homogenized system for the above fraction is:

$$\left(a_0' + a_1'x_k + \cdots + a_m'x_k^m\right) - f_k\left(1 + b_1'x_k + \cdots + b_n'x_k^n\right) = 0$$

where

$$k = 0: \qquad a'_0 + a'_1 x_0 + \cdots + a'_m x_0^m - b'_1 f_0 x_0 - \cdots - b'_n f_0 x_0^n = f_0$$

$$\vdots \qquad\qquad\qquad\qquad \vdots$$

$$k = m+n: \quad a'_0 + \cdots + a'_m x_{m+n}^m - b'_1 f_{m+n} x_{m+n} - \cdots - b'_n f_{m+n} x_{m+n}^n = f_{m+n}$$

The coefficient matrix for the system of the $m+n+1$ equations in $m+n+1$ parameters is:

$$A = \begin{bmatrix} 1 & x_0 & \cdots & x_0^m & x_0 f_0 & \cdots & x_0^n f_0 \\ 1 & x_1 & \cdots & x_1^m & x_1 f_1 & \cdots & x_1^n f_1 \\ \vdots & \vdots & \vdots & \vdots & \vdots & \vdots & \vdots \\ 1 & x_{m+n} & \cdots & x_{m+n}^m & x_{m+n} f_{m+n} & \cdots & x_{m+n}^n f_{m+n} \end{bmatrix}$$

As there is an answer for this system, we conclude that $|A| \neq 0$ and this determinant is the denominator of $\phi^{m,n}(x)$ for $x=0$, where $\phi^{m,n}(0) = a_0$ which is obtained using the Kramer's rule.

4.2.3 Problem

Calculate the Pade approximation of $\cos(x)$ for $m,n = 4$.
  Solution: We know

$$f(x) = \frac{P^m(x)}{Q^n(x)} = \frac{A + Mx + Bx^2 + Nx^3 + Cx^4}{1 + Lx + Dx^2 + Sx^3 + Ex^4}$$

Because $\cos(x)$ is an even function, we do not consider the odd powers. Thus, the five unknowns are remained, and as a result, we write five sentences of McLaurin series $f(x)$.

$$1 - \frac{x^2}{2!} + \frac{x^4}{4!} - \frac{x^6}{6!} + \frac{x^8}{8!} = \frac{A + Bx^2 + Cx^4}{1 + Dx^2 + Ex^4}$$

As a result:

$$A + Bx^2 + Cx^4 = 1 + \left(D - \frac{1}{2}\right)x^2 + \left(\frac{1}{4!} - \frac{D}{2} + E\right)x^4$$

$$+ \left(-\frac{1}{6!} + \frac{D}{4!} - \frac{E}{2}\right)x^6 + \left(\frac{1}{8!} - \frac{D}{6!} + \frac{E}{4!}\right)x^8$$

$$+ \left(\frac{D}{8!} - \frac{E}{6!}\right)x^{10} + \frac{E}{8!}x^{12}$$

Since $m + n = 8$, then we consider x up to the 8th power.

$$A = 1$$

$$B = D - \frac{1}{2}$$

$$C = \frac{1}{24} - \frac{D}{2} + E$$

$$\frac{D}{24} - \frac{E}{2} = \frac{1}{720}$$

$$-\frac{D}{720} + \frac{E}{24} = -\frac{1}{40320}$$

By calculating the unknown parameters, we obtain:

$$\cos x = \frac{1 - \dfrac{11}{252}x^2 + \dfrac{2567}{30240}x^4}{1 + \dfrac{11}{252}x^2 + \dfrac{13}{15120}x^4}$$

4.3 Inverse Newton's Divided Difference Interpolation

In this section, we introduce interpolations that are a special case of fractional and recursive interpolation. Among these interpolations are inverse differences, which are defined recursively for interpolation points (x_i, f_i), $i = 0, 1, \ldots$ as follows:

$$\varphi(x_i, x_j) = \frac{x_i - x_j}{f_i - f_j}$$

$$\varphi(x_i, x_j, x_k) = \frac{x_j - x_k}{\varphi(x_i, x_j) - \varphi(x_i, x_k)} \tag{4.7}$$

$$\varphi(x_i, \ldots, x_l, x_m, x_n) = \frac{x_m - x_n}{\varphi(x_i, \ldots, x_l, x_m) - \varphi(x_i, \ldots, x_l, x_n)}$$

The above process is an Aitken recursive interpolation process. Now using the equations of (4.7), we introduce a fractional expression that interpolates points (x_i, f_i), $i = 0, 1, \ldots$ as follows:

$$\phi^{n,n}(x) = f_0 + x - x_0 \left\lfloor \varphi(x_0, x_1) + x - x_1 \right\lfloor \varphi(x_0, x_1, x_2)$$
$$+ x - x_2 \left\lfloor \varphi(x_0, x_1, x_2, x_3) + \cdots + x - x_{2n-1} \right\lfloor \varphi(x_0, \ldots, x_{2n}) \right. \tag{4.8}$$

This fractional expression is called inverse difference interpolator. This method moves in the data table according to a right triangle and has no symmetry. Therefore, we introduce another method called reciprocal differences, which is defined recursively and is symmetrical.

$$\rho(x_i) = f_i$$

$$\rho(x_i, x_j) = \frac{x_i - x_j}{f_i - f_j} \tag{4.9}$$

$$\rho(x_i, x_{i+1}, \ldots, x_{i+k}) = \frac{x_i - x_{i+k}}{\rho(x_i, \ldots, x_{i+k-1}) - \rho(x_{i+1}, \ldots, x_{i+k}) + \rho(x_{i+1}, \ldots, x_{i+k-1})}$$

It can be easily shown that

$$\varphi(x_0, \ldots, x_p) = \rho(x_0, \ldots, x_p) = \rho(x_0, \ldots, x_{p-2}) \tag{4.10}$$

Using the equation (4.10), the fractional term of (4.8) is obtained as follows:

$$\phi^{n,n}(x) = f_0 + x - x_0 \left\lfloor \rho(x_0, x_1) + x - x_1 \right\lfloor \rho(x_0, x_1, x_2) - \rho(x_0)$$
$$+ \cdots + x - x_{2n-1} \left\lfloor \rho(x_0, \ldots, x_{2n}) - \rho(x_0, \ldots, x_{2n-2}) \right. \tag{4.11}$$

The fractional term of (4.11) is called the reciprocal differences interpolator for points (x_i, f_i), $i = 0, \ldots, n$.

4.3.1 Problem

Write the inverse and reciprocal differences for the below interpolation points and obtain the fractional function $\phi^{2,2}(x)$ such that the degree of nominator and denominator is 2 and $\phi^{2,2}(x_i) = f_i$ (Table 4.1).

TABLE 4.1

Interpolation Points for Problem 4.3.1

x_i	0	1	−1	2	−2
F_i	1	3	$\dfrac{3}{5}$	3	$\dfrac{3}{5}$

TABLE 4.2

Inverse Difference Method Points for Problem 4.3.1

x_i	f_i	$\varphi(x_0, x_i)$	$\varphi(x_0, x_1, x_i)$	$\varphi(x_0, x_1, x_2, x_i)$	$\varphi(x_0, x_1, x_2, x_3, x_i)$
0	1				
1	3	$\dfrac{1}{2}$			
−1	$\dfrac{3}{5}$	$\dfrac{5}{2}$	−1		
2	3	1	2	1	
−2	$\dfrac{3}{5}$	5	$\dfrac{-2}{3}$	−3	1

Solution: Inverse differences method

For Table 4.2, we have:

$$\varphi\left(x_i, x_j\right) = \frac{x_i - x_j}{f_i - f_j}$$

$$\varphi\left(x_i, \ldots, x_m, x_k, x_l\right) = \frac{x_k - x_l}{\varphi\left(x_i, \ldots, x_k\right) - \varphi\left(x_i, \ldots, x_m, x_l\right)}$$

So,

$$\phi^{2,2}\left(x\right) = f_0 + \cfrac{x - x_0}{\varphi\left(x_0, x_1\right) + \cfrac{x - x_1}{\varphi\left(x_0, x_1, x_2\right) + \cfrac{x - x_2}{\varphi\left(x_0, x_1, x_2, x_3\right) + \cfrac{x - x_3}{\varphi\left(x_0, \ldots, x_4\right)}}}}$$

$$= 1 + \cfrac{x}{\dfrac{1}{2} + \cfrac{x - 1}{-1\dfrac{x+1}{1 + \dfrac{x-2}{1}}}}$$

$$= \frac{x^2 + 2}{x^2 - 2x + 2}$$

Reciprocal differences method:

The table of inverse differences can be obtained using the following equations:

$$\rho(x_i, x_j) = \frac{x_i - x_j}{f_i - f_j}$$

$$\rho(x_i, x_{i+1}, ..., x_{i+k}) = \frac{x_i - x_{i+k}}{\rho(x_i, ..., x_{i+k-1}) - \rho(x_{i+1}, ..., x_{i+k})}$$
$$+ \rho(x_{i+1}, ..., x_{i+k-1})$$

So according to the equation (4.11), we have:

$$\phi^{2,2}(x) = \frac{x^2 + 2}{x^2 - 2x + 2}$$

Fractional method:

$$\phi^{2,2}(x_i) = \frac{a_2 x^2 + a_1 x + a_0}{b_2 x^2 + b_1 x + b_0}$$

At each of the interpolation points, we will have:

$$\phi^{2,2}(x_i) = \frac{a_2 x_i^2 + a_1 x_i + a_0}{b_2 x_i^2 + b_1 x_i + b_0} = f_i, \quad i = 0, ..., 4$$

From the above relation, it follows:

$$\left(a_2 x_i^2 + a_1 x_i + a_0 \right) - f \left(b_2 x_i^2 + b_1 x_i + b_0 \right) = 0, \quad i = 0, ..., 4$$

By assuming $a_0 = 2$ and solving the above 5×5 system, the above fraction is obtained.

4.3.2 Problem

Show that $\rho(x_1, ..., x_n)$ is invariant under permutation of indices.
    Solution: According to the equation (4.9), the continued fractions are as follows:

$$f(x_1) + \cfrac{x - x_1}{\rho(x_1, x_2) + \cfrac{x - x_2}{\rho_2(x_1, x_2, x_3) - f(x_1) + \cfrac{x - x_3}{\rho_3(x_1, ..., x_4) - \rho_1(x_1, x_2) + \cdots}}}$$

where

$$p_{n-1}(x_1,\ldots,x_n) = \frac{x_1 - x_n}{p_{n-2}(x_1,\ldots,x_{n-1}) - p_{n-2}(x_2,\ldots,x_n)} + p_{n-3}(x_2,\ldots,x_{n-1})$$

First case: for x_i, $i = 1,2,\ldots,2n$, the fractional interpolator is as $\frac{p(x)}{q(x)}$, where $p(x)$ is of degree n and $q(x)$ is of degree $n-1$, so they are written as follows:

$$p(x) = a_0 + a_1 x + \cdots + a_{n-1} x^{n-1} + x^n$$

$$q(x) = b_0 + b_1 x + \cdots + b_{n-2} x^{n-2} + x^{n-1} p_{2n-1}$$

How to write the above relation is clear due to the form of continued fractions. For example, if $n = 1$, $i = 1,2$, then

$$f(x_1) + \frac{x - x_1}{p_1(x_1,x_2)} = \frac{p_1(x_1,x_2) \cdot f(x_1) + (x - x_1)}{p_1(x_1,x_2)}$$

It is clear that the coefficient x in the nominator is equal to 1 and the coefficient x^0 in the denominator is equal to p_1. Now to approximate the function y, we write:

$$y = \frac{p(x)}{q(x)} + R(x)$$

where $R(x)$ is zero for all interpolation points of x_i, $i = 1,2,\ldots,2n$. We put

$$y_i = \frac{p(x_i)}{q(x_i)}, \qquad i = 1,\ldots,2n$$

$$\Rightarrow p(x_i) - y_i \cdot q(x_i) = 0, \qquad i = 1,\ldots,2n$$

By substituting the relations of $p(x)$ and $q(x)$ in term of x_i in the above relation we will have:

$$a_0 + a_1 x_i + \cdots + a_{n-1} x_i^{n-1} + x_i^n - b_0 y_i - b_1 x_i y_i - \cdots - x_i^{n-1} y_i p_{2n-1} = 0$$

where for $i = 1,\ldots,2n$, we will have $2n$ equations as above. The unknowns are $a_0,\ldots,a_{n-1}$ and $b_0,\ldots,b_{n-2}$, the number of which is $2n-1$, and if we consider p_{2n-1} as unknown parameters, the number of unknowns becomes equal to $2n$ and a square system is obtained. The above relation can be rearranged and x_i^n can be moved to the right side of the equation so that:

$$a_0 - b_0 y_i + a_1 x_i - b_1 x_i y_i + \cdots + a_{n-1} x_i^{n-1} - x_i^{n-1} y_i p_{2n-1} = -x_i^n$$

Now by using the Kramer's rule to get $\rho_{2n-1}(x_1, \ldots, x_{2n})$, we have:

$$\rho_{2n-1}(x_1, \ldots, x_{2n}) = \frac{\begin{vmatrix} 1 & y_i & x_i & x_i y_i & \cdots & x_i^{n-1} & x_i^n \end{vmatrix}_{i=1}^{2n}}{\begin{vmatrix} 1 & y_i & x_i & x_i y_i & \cdots & x_i^{n-1} & x_i^{n-1} y_i \end{vmatrix}_{i=1}^{2n}}$$

where the nominator and denominator determinants have $2n$ rows for $i = 1, \ldots, 2n$. Now, if we exchange two points, only two rows of the nominator and denominator determinants are exchanged so that the value of the determinants does not change and only the nominator and the denominator are multiplied by a minus sign, and finally, the value of ρ_{2n-1} remains unchanged.

Second case: for x_i, $i = 1, \ldots, 2n+1$, fractional interpolators are as $\frac{p(x)}{q(x)}$, where $p(x)$ and $q(x)$ are both of degree n and are written as follows:

$$p(x) = c_0 + c_1 x + \cdots + c_{n-1} x^{n-1} + x^n \rho_{2n}$$

$$q(x) = d_0 + d_1 x + \cdots + d_{n-1} x^{n-1} + x^n$$

As in the first case, the above relations can be written using continued fractions. For example, if $n = 1$ and $i = 1, 2, 3$, then

$$f(x_1) = \cfrac{x - x_1}{\rho_1(x_1, x_2) + \cfrac{x - x_2}{\rho_2(x_1, x_2, x_3) - f(x_1)}}$$

After simplification, the coefficient of x^1 in the nominator is equal to ρ_2 and the coefficient of x^1 in the denominator is equal to 1. Now, in order to approximate y, we have:

$$y = \frac{p(x)}{q(x)} + R(x)$$

where $R(x)$ is zero for interpolation points of x_i, $i = 1, \ldots, 2n+1$. So

$$y_i = \frac{p(x_i)}{q(x_i)}, \quad i = 1, \ldots, 2n+1$$

Therefore,

$$p(x_i) - y_i \cdot q(x_i) = 0, \quad i = 1, \ldots, 2n+1$$

Now by substituting $p(x)$ and $q(x)$ in terms of x_i in the above relation and some simplification, we have:

$$c_0 - d_0 y_i + c_1 x_i - d_1 x_i y_i + \cdots + c_{n-1} x_i^{n-1} - d_{n-1} x_i^{n-1} y_i - x_i^n y_i + x_i^n \rho_{2n} = 0$$

As for first case, we have $2n + 1$ equation and if we consider ρ_{2n} as unknown parameters, we have a square system. As $x_i^n y_i$ is the known value of equation, so:

$$c_0 - d_0 y_i + c_1 x_i - d_1 x_i y_i + \cdots + c_{n-1} x_i^{n-1} - d_{n-1} x_i^{n-1} y_i + x_i^n \rho_{2n} = x_i^n y_i$$

By using the Kramer's rule for calculating ρ_{2n}, we obtain:

$$\rho_{2n}(x_1, \ldots, x_{2n+1}) = \frac{\begin{vmatrix} 1 & y_i & x_i & x_i y_i & \cdots & x_i^{n-1} y_i & x_i^n y_i \end{vmatrix}_{i=1}^{2n+1}}{\begin{vmatrix} 1 & y_i & x_i & x_i y_i & \cdots & x_i^{n-1} y_i & x_i^{n-1} \end{vmatrix}_{i=1}^{2n+1}}$$

As in the first case, it is clear that by exchanging the position of the two points, the position of the two rows in the nominator and the denominator exchanges, and finally, the value of ρ_{2n} remains unchanged.

4.3.3 Problem

Show that inverse differences are not invariant under permutation of indices.
   Solution: we have:

$$(x_j, x_k, x_l) = \frac{x_k - x_l}{\phi(x_j, x_k) - \phi(x_j, x_l)}$$

$$= \frac{x_k - x_l}{\dfrac{x_j - x_k}{f_j - f_k} - \dfrac{x_j - x_l}{f_j - f_l}}$$

$$= \frac{(x_k - x_l)(f_j - f_k)(f_j - f_l)}{(x_j - x_k)(f_j - f_l) - (x_j - x_l)(f_j - f_k)}$$

$$\neq \phi(x_l, x_k, x_j)$$

4.4 Trigonometric Interpolation

When the function f is a periodic function, we approximate it with periodic functions that have a definite periodicity. Therefore, in this section, we examine the trigonometric interpolating functions. This type of interpolation is a special mode of linear and non-recursive interpolation. The structure of a trigonometric interpolator is a combination of functions $\sin lx$ and $\cos lx$ where $l \in \mathbb{Z}$.

Suppose that for $k = 0, \ldots, N-1$, (x_k, f_k) s are interpolation points. We consider two modes for the trigonometric interpolating function:

$$
\psi(x) = \begin{cases}
\dfrac{A_0}{2} + \displaystyle\sum_{l=1}^{M} (A_l \cos lx + B_l \sin lx), & N = 2M+1 \\[4mm]
\dfrac{A_0}{2} + \displaystyle\sum_{l=1}^{M-1} (A_l \cos lx + B_l \sin lx) + \dfrac{A_M}{2} \cos Mx, & N = 2M
\end{cases}
$$

In both cases, $\psi(x)$ is a periodic function in terms of x with periodicity of 2π.

Suppose that the interpolation points are as $0 = x_0 < x_1 < \cdots < x_{N-1} = 2\pi$ where

$$
x_k = \frac{2k\pi}{N}, \quad k = 0, \ldots, N-1
$$

Then, we can write:

$$
e^{-ilx_k} = e^{\frac{-2k\pi i l}{N}} = e^{\frac{2ki\pi(N-l)}{N}} = e^{(N-l)ix_k}
$$

$$
\cos lx_k = \frac{e^{ilx_k} + e^{(N-l)ix_k}}{2}, \quad \sin lx_k = \frac{e^{ilx_k} - e^{(N-l)ix_k}}{2i}
$$

(4.12)

Using the equation (4.12), the functions $\psi(x)$ are converted to the function $p(x)$ with the complex coefficients as follows:

$$
p(x) = \beta_0 + \beta_1 e^{ix} + \cdots + \beta_{N-1} e^{(N-1)ix}
$$

which is called phase or base polynomials. Given that the set $\left\{1, e^{ix}, \ldots, e^{(N-1)ix}\right\}$ is a basis for space, so polynomial $p(x)$ is unique. Now we find the coefficients β_j so that the following interpolation condition holds:

$$
\psi(x_k) = p(x_k) = f_k, \quad k = 0, \ldots, N-1
$$

4.4.1 Problem

Phase polynomials of $p(x) = \displaystyle\sum_{j=0}^{N-1} \beta_j \cdot e^{jix}$ where f_k is a complex number and $x_k = \dfrac{2k\pi}{N}$, satisfy the interpolation condition of $p(x_k) = f_k$, $k = 0, \ldots, N-1$, if and only if

$$\beta_j = \frac{1}{N}\sum_{k=0}^{N-1}f_k w_k^{-j}$$

where $w_k^j = e^{jix_k}$

Solution: Proof of necessity: If we consider the N-dimensional vectors of F and W^j as follows:

$$F = \left(f_0,\ldots,f_{N-1}\right)^t, \quad W^j = \left(w_0^j,\ldots,w_{N-1}^j\right)^t$$

As for $k = 0,\ldots,N-1$, $p(x_k) = f_k$, then

$$\sum_{r=0}^{N-1}\beta_r \cdot e^{rix_k} = f_k$$

So,

$$\sum_{r=0}^{N-1}\beta_r w_k^r = f_k$$

As a result

$$\sum_{r=0}^{N-1}\beta_r w^r = F$$

So, we will have

$$\frac{1}{N}\sum_{r=0}^{N-1}f_k w_k^{-j} = \frac{1}{N}\left\langle F, W^j \right\rangle$$

$$= \frac{1}{N}\left\langle \sum_{r=0}^{N-1}\beta_r W^r, W^j \right\rangle$$

$$= \frac{1}{N}\sum_{r=0}^{N-1}\beta_r \left\langle W^r, W^j \right\rangle$$

$$= \frac{1}{N} \cdot N\beta_j$$

$$= \beta_j$$

Proof of adequacy:

$$p(x) = \sum_{r=0}^{N-1} \beta_j \cdot e^{jix} = \sum_{r=0}^{N-1} \beta_j w^j = p(w)$$

So,

$$p(w_l) = \sum_{j=0}^{N-1} \beta_j w_l^j$$

$$= \frac{1}{N} \sum_{j=0}^{N-1} \sum_{k=0}^{N-1} f_k w_k^{-j} w_l^j$$

$$= \frac{1}{N} \sum_{j=0}^{N-1} \sum_{k=0}^{N-1} f_k w_j^{-k} w_j^l$$

$$= \frac{1}{N} \sum_{k=0}^{N-1} f_k \sum_{j=0}^{N-1} w_j^{-k} w_j^l$$

$$= \frac{1}{N} f_l \cdot N$$

$$= f_l$$

Then, $p(w_l) = f_l$ and the statement is true.

4.4.2 Problem

For interpolation points of $\{(x_i, f_i)\}_{i=0}^{n-1}$ that are periodic, the interpolation polynomials are as $p \in \Pi_{n-1}$ and $p(x_i) = f_i$, $i = 0,1,\ldots,n-1$ where p is any of the arbitrary interpolation polynomials. State reasonably that which polynomial is the best approximation for these points.

Solution: s-pieces polynomials of p_s which obtained from the phase interpolation polynomials of $p(x)$ are the best approximation for the above points. Because if function $q(x)$ is as follows:

$$q(x) = \gamma_0 + \gamma_1 e^{ix} + \cdots + \gamma_s e^{six}$$

We can prove that $S(p_s) \le S(q)$, where

$$S(q) = \sum_{k=0}^{n-1} |f_k - q(x_k)|^2$$

We introduce the n-dimensional vectors of p_s and q as follows:

$$p_s = \left(p_s(x_0),\ldots,p_s(x_{n-1})\right), \quad q = \left(q(x_0),\ldots,q(x_{n-1})\right)$$

$S(q)$ can be written as an inner product of $S(q) = \langle f - q, f - q \rangle$.

$$\|f - q\|_2^2 = \sum_{k=0}^{n-1} |f_k - q(x_k)|^2 = \langle f - q, f - q \rangle$$

Given that for $j = 0,\ldots,n-1, \langle f, w^j \rangle = N\beta_j$ so for every j in $j \le s$, we have:

$$\langle f - p_s, w^j \rangle = \left\langle f - \sum_{h=0}^{s} \beta_h w^h, w^j \right\rangle$$

$$= \langle f, w^j \rangle - \left\langle \sum_{h=0}^{s} \beta_h w^h, w^j \right\rangle$$

$$= \langle f, w^j \rangle - \sum_{h=0}^{s} \beta_h \langle w^h, w^j \rangle$$

$$= N\beta_j - N\beta_j$$

$$= 0$$

$$\langle f - p_s, p_s - q \rangle = \left\langle f - p_s, \sum_{j=0}^{s} \beta_j w^j - \sum_{j=0}^{s} \gamma_j w^j \right\rangle$$

$$= \left\langle f - p_s, \sum_{j=0}^{s} (\beta_j - \gamma_j) w^j \right\rangle$$

$$= \sum_{j=0}^{s} (\beta_j - \gamma_j)\langle f - p_s, w^j \rangle$$

$$= 0$$

So, finally we will have:

$$S(q) = \langle f - q, f - q \rangle$$
$$= \langle (f - p_s) + (p_s - q), (f - p_s) + (p_s - q) \rangle$$
$$\geq \langle f - p_s, f - p_s \rangle$$
$$= S(p_s)$$

4.4.3 Problem

A) Show that for points with lengths of x_k, $a \leq x_0 < x_1 < \cdots < x_{2n} < a + 2\pi$ and widths of $y_0, \ldots, y_{2n}$, there exists a unique trigonometric polynomial as

$$T(x) = \frac{1}{2} a_0 + \sum_{j=1}^{n} \left(a_j \cos jx + b_j \sin jx \right)$$

with property of $T(x_k) = y_k$, $k = 0, \ldots, 2n$.

B) If $y_0, \ldots, y_{2n}$ are real numbers, prove that the coefficients a_j and b_j will also be real.

Solution: A) we partition the interval $[a, a + 2\pi]$ as follows:

$$t_i = a + i \frac{\pi}{n}, \quad i = 0, \ldots, 2n$$

So, we can say that

$$e^{-jia} = e^{ia(2n-j)}$$

(Why?) and as a result

$$e^{-jit_k} = e^{-ji\left(a + \frac{2k\pi}{2n}\right)}$$

$$= e^{-jia} \cdot e^{-ji\frac{2k\pi}{2n}}$$

$$= e^{ia(2n-j)} \cdot e^{-ji\frac{2k\pi}{2n}} \cdot e^{2k\pi i \frac{2n}{2n}}$$

$$= e^{ia(2n-j)} \cdot e^{i(2n-j)\frac{2k\pi}{2n}}$$

$$= e^{i(2n-j)\left(a + \frac{2k\pi}{2n}\right)}$$

$$= e^{i(2n-j)t_k}$$

Without the loss of generality, we assume that for every x_k of $[a, a+2\pi]$, t_i and t_{i+1} are the closest points from left and right to x_k, so that

$$x_k = \lambda t_i + (1-\lambda) t_{i+1}, \quad \lambda \in [0,1]$$

We prove that

$$e^{-jix_k} = e^{(2n-j)ix_k}$$

For this purpose, we have:

$$e^{-jix_k} = e^{-ji(\lambda t_i + (1-\lambda) t_{i+1}}$$

$$= e^{-ji\lambda t_i} \cdot e^{-ji(1-\lambda) t_{i+1}}$$

$$= e^{\lambda(2n-j)it_i} \cdot e^{(1-\lambda)(2n-j)it_{i+1}}$$

$$= e^{(2n-j)i(\lambda t_i + (1-\lambda) t_{i+1})}$$

$$= e^{(2n-j)ix_k}$$

Now we prove that $T(x)$ is unique. By substituting $\cos jx = \dfrac{e^{ijx} + e^{-ijx}}{2}$ and $\sin jx = \dfrac{e^{ijx} - e^{-ijx}}{2i}$ in $T(x)$, we have:

$$T(x) = \frac{1}{2} a_0 + \sum_{j=1}^{n} \left(a_j \cos jx + b_j \sin jx \right)$$

$$= \frac{1}{2} a_0 + \sum_{j=1}^{n} \left[\frac{1}{2} \left(a_j e^{ijx} + a_j e^{-ijx} \right) + \frac{1}{2} \left(-ib_j e^{ijx} + ib_j e^{-ijx} \right) \right]$$

$$= \frac{1}{2} a_0 + \left[\frac{1}{2} \sum_{j=1}^{n} \left[(a_j - ib_j) \right] e^{ijx} \left(a_j + ib_j \right) \right] e^{-ijx}$$

$$= \frac{1}{2} a_0 + \left[\frac{1}{2} \sum_{j=1}^{n} \left[(a_j - ib_j) \right] e^{ijx} + \left(a_j + ib_j \right) e^{(2n-j)ix} \right]$$

By simplifying the above relation, $T(x)$ is obtained as follows:

$$T(x) = \beta_0 + \beta_1 e^{ix} + \beta_2 e^{2ix} + \cdots + \beta_{2n-1} e^{(2n-1)ix}$$

which is a unique polynomial. Therefore, $T(x)$ is also unique.

B) According to part A, we have:

$$T(x) = \frac{1}{2}a_0 + \frac{1}{2}\sum_{j=1}^{n}\left[\left(a_j - ib_j\right)e^{ijx} + \left(a_j + ib_j\right)e^{-ijx}\right]$$

$$= \frac{1}{2}a_0 + \frac{1}{2}\sum_{j=1}^{n}\left[\left(a_j - ib_j\right)\right]e^{ijx} + \frac{1}{2}\sum_{j=1}^{n}\left(a_j + ib_j\right)e^{-ijx}$$

$$= \frac{1}{2}a_0 + \frac{1}{2}\sum_{j=1}^{n}\left[\left(a_j - ib_j\right)\right]e^{ijx} + \frac{1}{2}\sum_{j=-n}^{-1}\left(a_{-j} + ib_{-j}\right)e^{ijx}$$

We put

$$c_j = \begin{cases} \frac{1}{2}\left(a_j - ib_j\right), & 1 \le j \le n \\ \frac{1}{2}\left(a_{-j} + ib_{-j}\right), & -n \le j \le -1 \\ \frac{a_0}{2}, & j = 0 \end{cases}$$

In this case, it can be written as

$$T(x) = \sum_{j=-n}^{n} c_j e^{ijx}$$

Now we prove that $c_{-j} = \bar{c}_j$:

$$c_{-j} = \begin{cases} \frac{1}{2}\left(a_{-j} - ib_{-j}\right), & 1 \le -j \le n \\ \frac{1}{2}\left(a_j + ib_j\right), & -n \le -j \le -1 \\ \frac{a_0}{2}, & j = 0 \end{cases}$$

$$= \begin{cases} \frac{1}{2}\left(a_j - ib_j\right), & -n \le j \le -1 \\ \frac{1}{2}\left(a_{-j} + ib_{-j}\right), & 1 \le j \le n \\ \frac{a_0}{2}, & j = 0 \end{cases}$$

$$= \bar{c}_j$$

Therefore, a_j and b_j are real.

4.4.4 Problem

a) Show that for real values of $x_1,\ldots,x_{n2n}$,

$$t(x) = \prod_{k=1}^{2n} \sin\left(\frac{x - x_k}{2}\right)$$

is a trigonometric polynomial in the form of:

$$\frac{1}{2}a_0 + \sum_{j=1}^{n}\left(a_j \cos jx + b_j \sin jx\right)$$

with real coefficients of a_j and b_j.

b) Prove that the trigonometric interpolating polynomial, for points of lengths x_k where

$$a \le x_0 < \cdots < x_{2n} < a + 2\pi$$

and widths $y_0,\ldots,y_{2n}$, is a polynomial of

$$T(x) = \sum_{j=0}^{2n} y_j t_j(x)$$

in which $t_j(x)$ is defined as:

$$t_j(x) = \frac{\displaystyle\prod_{k=0,k\neq j}^{2n} \sin\left(\frac{x - x_k}{2}\right)}{\displaystyle\prod_{k=0,k\neq j}^{2n} \sin\left(\frac{x_j - x_k}{2}\right)}$$

Solution: a) Trigonometric expression of

$$p(x) = \frac{1}{2}a_0 + \sum_{j=1}^{n}\left(a_j coxjx + b_j \sin jx\right)$$

according to the relationship between de Moivre's formula and uniform partition $x_k = \dfrac{2k\pi}{2n}, k = 0,\ldots,2n-1$, on the interval $[0,2\pi]$ can be written as the following problem

$$p(x) = \beta_0 + \beta_1 e^{ix} + \cdots + \beta_{2n-1}e^{(2n-1)ix}$$

where β_j is complex and $p(x_k) = f_k$. So, we prove that

$$t(x) = \prod_{k=1}^{2n} \sin\left(\frac{x - x_k}{2}\right) = \beta_0 + \beta_1 e^{ix} + \cdots + \beta_{2n-1} e^{(2n-1)ix}$$

We prove the above e by induction on n.

Initial case: if $n = 1$, we have:

$$t(x) = \prod_{k=1}^{2} \sin\left(\frac{x - x_k}{2}\right)$$

$$= \sin\left(\frac{x - x_1}{2}\right)\sin\left(\frac{x - x_2}{2}\right)$$

$$= \left(\frac{1}{2i}\right)^2 \left[e^{i\left(\frac{x-x_1}{2}\right)} - e^{-i\left(\frac{x-x_1}{2}\right)}\right]\left[e^{i\left(\frac{x-x_2}{2}\right)} - e^{-i\left(\frac{x-x_2}{2}\right)}\right]$$

$$= \frac{-1}{4}\left[e^{ix - \frac{i}{2}(x_1 + x_2)} - e^{-\frac{i}{2}(x_1 - x_2)} - e^{\frac{i}{2}(x_1 - x_2)} + e^{-ix + \frac{i}{2}(x_1 + x_2)}\right]$$

$$= \frac{-1}{4}(\cos x + i \sin x)\left(\cos\frac{x_1 + x_2}{2} - i \sin\frac{x_1 + x_2}{2}\right)$$

$$+ \frac{1}{4}\left(\cos\frac{x_1 - x_2}{2} - i \sin\frac{x_1 - x_2}{2}\right) + \frac{1}{4}\left(\cos\frac{x_1 - x_2}{2} + i \sin\frac{x_1 - x_2}{2}\right)$$

$$- \frac{1}{4}(\cos x - i \sin x)\left(\cos\frac{x_1 + x_2}{2} + i \sin\frac{x_1 + x_2}{2}\right)$$

$$= \frac{1}{2}\cos\left(\frac{x_1 - x_2}{2}\right) - \frac{1}{2}\cos\left(\frac{x_1 + x_2}{2}\right) \cdot \cos x - \frac{1}{2}\sin\left(\frac{x_1 + x_2}{2}\right) \cdot \sin x$$

$$= \frac{a_0}{2} + a_1 \cos x + b_1 \sin x$$

where

$$a_0 = \cos\frac{x_1 - x_2}{2}, \quad a_1 = \frac{-1}{2}\cos\frac{x_1 + x_2}{2}, \quad b_1 = \frac{-1}{2}\sin\frac{x_1 + x_2}{2}$$

Induction hypothesis: Suppose that for $n > 1$, we have:

$$t(x) = \prod_{k=1}^{2n} \sin\left(\frac{x - x_k}{2}\right) = \beta_0 + \beta_1 e^{ix} + \cdots + \beta_{2n-1} e^{(2n-1)ix}$$

Induction step: We prove for $n+1$ that:

$$t(x) = \prod_{k=1}^{2n+2} \sin\left(\frac{x - x_k}{2}\right) = \beta_0 + \beta_1 e^{ix} + \cdots + \beta_{2n+1} e^{(2n+1)ix}$$

We will prove it as following:

$$t(x) = \prod_{k=1}^{2n+2} \sin\left(\frac{x - x_k}{2}\right)$$

$$= \prod_{k=1}^{2n} \sin\left(\frac{x - x_k}{2}\right) \cdot \sin\left(\frac{x - x_{2n+1}}{2}\right) \cdot \sin\left(\frac{x - x_{2n+2}}{2}\right)$$

$$= \beta_0 + \beta_1 e^{ix} + \cdots + \beta_{2n-1} e^{(2n-1)ix} \cdot \sin\left(\frac{x - x_{2n+1}}{2}\right) \cdot \sin\left(\frac{x - x_{2n+2}}{2}\right)$$

$$= \left(\beta_0 + \beta_1 e^{ix} + \cdots + \beta_{2n-1} e^{(2n-1)ix}\right) \cdot \left(\frac{1}{2i}\right)^2$$

$$\times \left[e^{\frac{i}{2}(x - x_{2n+1})} - e^{-\frac{i}{2}(x - x_{2n+1})}\right]\left[e^{\frac{i}{2}(x - x_{2n+2})} - e^{-\frac{i}{2}(x - x_{2n+2})}\right]$$

$$= \left(\beta_0 + \beta_1 e^{ix} + \cdots + \beta_{2n-1} e^{(2n-1)ix}\right)\left(\alpha_0 + \alpha_1 e^{ix} + \alpha_2 e^{-ix}\right)$$

$$= \beta_0\alpha_0 + \beta_0\alpha_1 e^{ix} + \beta_0\alpha_2 e^{-ix} + \beta_1\alpha_0 e^{ix} + \beta_1\alpha_1 e^{2ix} + \beta_1\alpha_2 + \cdots$$

$$+ \beta_{2n-1}\alpha_0 e^{(2n-1)ix} + \beta_{2n-1}\alpha_1 e^{2nix} + \beta_{2n-1}\alpha_2 e^{(2n-1)ix}$$

$$= \beta_0' + \beta_1' e^{ix} + \cdots + \beta_{2n+1}' e^{(2n+1)ix}$$

In the above relation, $e^{-ix} = e^{[(2n+2)-1]ix}$ is assumed.
b) We must prove $T(x_i) = y_i$ for $i = 0, \ldots, 2n$.
According to the definition of $t_j(x)$, we have $t_j(x_i) = \delta_{ji}$, because if $i \neq j$, then:

$$t_j(x_i) = \frac{\displaystyle\prod_{k=0, k \neq j}^{2n} \sin\left(\frac{x_i - x_k}{2}\right)}{\displaystyle\prod_{k=0, k \neq j}^{2n} \sin\left(\frac{x_j - x_k}{2}\right)}$$

Thus,

$$T(x_i) = \sum_{j=0}^{2n} y_j t_j(x_0) = y_i$$

4.4.5 Problem

a) Show for the integer j that

$$\sum_{k=0}^{2m} \cos jx_k = (2m+1)h(j)$$

$$\sum_{k=0}^{2m} \sin jx_k = 0$$

where

$$x_k := \frac{2\pi k}{2m+1}$$

$$h(j) := \begin{cases} 1, & j = 0 \bmod 2m+1 \\ 0, & \text{o.w.} \end{cases}$$

b) Using (a), derive the following relations for j and k:

$$\sum_{i=0}^{2m} \sin jx_i \cdot \sin kx_i = \frac{2m+1}{2}\left[h(j-k) - h(j+k)\right]$$

$$\sum_{i=0}^{2m} \cos jx_i \cdot \cos kx_i = \frac{2m+1}{2}\left[h(j-k) - h(j+k)\right]$$

$$\sum_{i=0}^{2m} \cos jx_i \cdot \sin kx_i = 0$$

Solution: a) It should be proved that:

$$\sum_{k=0}^{2m} \cos jx_k = \sum_{k=0}^{2m} \cos\left(j\frac{2k\pi}{2m+1}\right) = \begin{cases} 2m+1, & j = 0 \bmod 2m+1 \\ 0, & \text{o.w.} \end{cases}$$

If $j = (2m+1)q$ with $q \in \mathbb{Z}$ then,

$$\sum_{k=0}^{2m} \cos\left(j\frac{2k\pi}{2m+1}\right) = \sum_{k=0}^{2m} \cos(2k\pi q) = \sum_{k=0}^{2m} 1 = 2m+1$$

If $j = (2m+1)q + r$ with $0 < r \le 2m$, then

$$\sum_{k=0}^{2m} \cos\left(j\frac{2k\pi}{2m+1}\right) = \sum_{k=0}^{2m} \cos\left[((2m+1)q+r)\left(\frac{2k\pi}{2m+1}\right)\right]$$

$$= \sum_{k=0}^{2m} \cos\left[2k\pi q + \frac{2k\pi r}{2m+1}\right]$$

$$= \sum_{k=0}^{2m} \cos\left(\frac{2k\pi r}{2m+1}\right)$$

$$= 1 + \cos\alpha + \cos 2\alpha + \cdots + \cos 2m\alpha$$

$$= \sum_{k=0}^{2m} \cos k\alpha$$

where $\alpha = \dfrac{2\pi r}{2m+1}$.

We know that:

$$\begin{cases} \cos\alpha + \cos 3\alpha + \cdots + \cos(2m-1)\alpha = \dfrac{\sin 2m\alpha}{2\sin\alpha} \\[3mm] \cos 0 + \cos 2\alpha + \cdots + \cos 2m\alpha = \dfrac{\cos m\alpha \cdot \sin(m+1)\alpha}{\sin\alpha} \end{cases} \qquad (4.13)$$

According to the equation (4.13), we have:

$$\sum_{k=0}^{2m} \cos k\alpha = \frac{\sin 2m\alpha + 2\cos m\alpha \cdot \sin(m+1)\alpha}{2\sin\alpha}$$

$$= \frac{2\sin m\alpha \cdot \cos m\alpha + 2\cos m\alpha \cdot \sin(m+1)\alpha}{2\sin\alpha}$$

$$= \frac{2\cos m\alpha \cdot 2\sin\left(\frac{2m+1}{2}\right)\alpha \cdot \cos\frac{\alpha}{2}}{2\sin\alpha}$$

$$= 0$$

Because

$$\sin\left(\frac{2m+1}{2}\right)\alpha = \sin \pi r = 0$$

Now if $j = (2m+1)q$, then

$$\sum_{k=0}^{2m} \sin jx_k = \sum_{k=0}^{2m} \sin\left(j\frac{2k\pi}{2m+1}\right)$$

$$= 0 + \sin\left(\frac{2\pi j}{2m+1}\right) + \cdots + \sin\left(\frac{2\pi(2m)j}{2m+1}\right)$$

$$= \sin 2\pi q + \cdots + \sin 4\pi mq$$

$$= 0$$

and if $j = (2m+1)q + r$ for $0 < r \le 2m$, and assuming $\alpha = \dfrac{2\pi j}{2m+1}$, we have:

$$\sum_{k=0}^{2m} \sin\left(\frac{2k\pi j}{2m+1}\right) = \sin\alpha + \sin 2\alpha + \cdots + \sin 2m\alpha$$

$$= \frac{\sin\left[\alpha + (2m)\dfrac{\alpha}{2} \cdot \sin\left(\dfrac{2m+1}{2}\alpha\right)\right]}{\sin\dfrac{\alpha}{2}}$$

$$= 0$$

Because

$$\sin\left(\frac{2m+1}{2}\right)\alpha = \sin \pi j = 0$$

$$\sin\alpha + \sin(\alpha + \beta) + \cdots + \sin(\alpha + (n-1)\beta)$$

$$= \frac{\sin\left(\alpha + (n-1)\dfrac{\beta}{2}\right) \cdot \sin\left(\dfrac{n\beta}{2}\right)}{\sin\dfrac{\beta}{2}}$$

b)

$$\sum_{i=0}^{2m} \sin jx_i \cdot \sin kx_i = \sum_{i=0}^{2m} \frac{-1}{2}\left[\cos\left(j+k\right)x_i - \cos\left(j-k\right)x_i\right]$$

$$= \frac{-1}{2}\sum_{i=0}^{2m}\cos\left(j+k\right)x_i + \frac{1}{2}\sum_{i=0}^{2m}\cos\left(j-k\right)x_i$$

$$= \frac{-1}{2}(2m+1)h\left(j+k\right) + \frac{1}{2}(2m+1)h\left(j-k\right) \quad \text{according to (a)}$$

$$= \frac{2m+1}{2}\left[h\left(j-k\right) - h\left(j+k\right)\right]$$

$$\sum_{i=0}^{2m} \cos jx_i \cdot \cos kx_i = \frac{1}{2}\sum_{i=0}^{2m}\left[\cos\left(j+k\right)x_i + \cos\left(j-k\right)x_i\right]$$

$$= \frac{1}{2}\sum_{i=0}^{2m}\cos\left(j+k\right)x_i + \frac{1}{2}\sum_{i=0}^{2m}\cos\left(j-k\right)x_i$$

$$= \frac{1}{2}(2m+1)h\left(j+k\right) + \frac{1}{2}(2m+1)h\left(j-k\right) \quad \left(\text{according to (a)}\right)$$

$$= \frac{2m+1}{2}\left[h\left(j-k\right) + h\left(j+k\right)\right]$$

$$\sum_{i=0}^{2m} \cos jx_i \cdot \sin kx_i = \frac{1}{2}\sum_{i=0}^{2m}\left[\sin\left(j+k\right)x_i - \sin\left(j-k\right)x_i\right]$$

$$= 0 \quad \left(\text{according to (a)}\right)$$

4.5 Spline Interpolation

Splines were first introduced in 1946 by a person named *I. J. Schoenbery*. If we want to describe this type of interpolator, we must say that it is one of the best interpolators considering the properties and characteristics of interpolation that will be discussed, because it is both a best approximation and a most accurate approximation. The best in the sense that the behavior of the interpolation points is approximated very smoothly, that is, it can be said that it is a functional approximation. Also, it is a most accurate approximation in the sense that in different points, it has approximations with much less error. So, it can be claimed that splines approximate the behavior of the function more accurately. In fact, there are two reasons why this interpolator is superior to other interpolators. First, in addition to interpolating points $(x_i, f_i), (x_i, f_i')$, and also (x_i, f_i''), this function can be said to approximate higher order derivatives, too. Second, this interpolator has a uniform convergence. As we have already seen, other interpolators did not have such properties. Although it can be claimed that the Hermite-type interpolation also has these two properties, but with a difference that in Hermite interpolation, points must exist in the form of $(x_i, f_i^{(k)})$. This means that consecutive derivatives of the function muse exist as the width of the points, and this is a disadvantage, because this is not always possible. On the other hand, spline works with the points (x_i, f_i) and does not need Hermite interpolation points, but interpolating convergence to the function and interpolating derivatives to the function derivatives are made in the spline conditions. Therefore, it can be said that if Hermite interpolation conditions are satisfied, Hermite interpolator has both of the mentioned important properties of the spline interpolating function. So, it can be said that spline is preferable to Hermite. Due to the good features mentioned about spline, many scientific applications can be enumerated for it, for example civil works, surveying, construction of international airport runway, medical issues, sonography, Farsi writing software, *Auto cad* engineering software, etc.

The appearance of the spline interpolating function is as a polynomial stepwise function. That is, it is a function of a number of rules, each of which is a polynomial. The same is true for the Hermite interpolating function. For example, if we want to examine the rule of a function of degree 3 (cubic) given the partition we have for the domain of the points, we have a polynomial of at most degree 3 on each of the subsets. Namely, it may even be a degree 3 spline with rules of degree 2 and/or 1, it may not even be of degree 3, and this means that these rules of degree 2 and/or 1 are intervals of polynomials of degree 3 on the subintervals.

Before examining spline interpolator, we first discuss about spline polynomials because in interpolation, interpolating space is always discussed and this space has bases: for example, in Lagrange interpolation which is a special type of polynomial interpolation, Lagrange polynomials form a basis of space, which is the case in other types as well. Regarding spline interpolation, it can be said that this interpolator is a linear combination of spline polynomials such as $S_i(x)$, which is introduced in the following:

$$p(x) = \sum_{i=1}^{n} c_i S_i(x)$$

Once $S_i(x)$ is known, it suffices to obtain c_is. In this case, the interpolation polynomials are discovered. For this purpose, we first discuss the space consisting of spline polynomials.

4.5.1 Spline Space

Suppose that $\Omega_n = \{x_i\}_{i=0}^n$ where $a = x_0 < x_1 < \cdots < x_n = b$ is a partition of the interval $[a,b] \subset \mathbb{R}$. For $i = 1,\ldots,n-1$, points x_i are usually called inner nodal points and points x_0 and x_n are end or border points.

4.5.2 Definition-Spline Polynomial Function

Function $S : [a,b] \to \mathbb{R}$ of degree $0 \leq l$ is called a spline polynomial function if it has the following conditions:

1. All derivatives of function S up to order $l-1$ exist and are continuous on the interval $[a,b]$.
2. Function S in each of the subsets $[x_i, x_{i+1})$ is a polynomial of at most degree l.

Assume that $S_l(\Omega_n)$ is the space of all spline polynomial functions of at most degree l on the partition Ω_2. It can be claimed that any polynomial of at most degree l is a spline in space $S_l(\Omega_n)$. But the opposite is not true.

4.5.3 Example

One-way function $q_{lv} : [a,b] \to \mathbb{R}, 0 \leq v \leq n-1$, which is defined as follows:

$$q_{lv}(x) = (x - x_v)_+^L = \begin{cases} (x - x_v)^l, & x \geq x_v \\ 0, & x < x_v \end{cases}$$

satisfies the two conditions of the definition (4.5.2), so $q_{lv} \in S_l(\Omega_n)$. These functions are called one-way splines.

Now that we are talking about spline space, we need to introduce a basis for it. Given the conditions of the definition (4.5.2), it is obvious that the basis members of this space all are polynomials of at most degree l, some of which can be considered as polynomials x^l and some as q_{lv}.

4.5.4 Definition

The basis of space $S_l(\Omega_n)$ is as

$$\left\{ 1, x, \ldots, x^l, q_{l1}, \ldots, q_{l,n-1} \right\}$$

so, the dimension of this space is $n + l$.

As mentioned earlier, splines are the best approximation among interpolation functions, so we will investigate whether it is the best approximation in spline space. For this purpose, we need a norm, so we inevitably require a normed space that includes the space of splines. Suppose that V is a normed linear space containing $S_l(\Omega_n)$. For example, V could be such a space:

$$\left(c[a,b],\|.\|_\infty\right), \qquad \left(c[a,b],\|.\|_2\right)$$

Now three things need to be considered:

1. Is the best approximation available on space V?
2. Why is it the best approximation?
3. Is the best approximation unique?

To answer the above questions, we need to discuss the best approximation.

4.5.5 Approximation

Suppose that $\left(V,\|.\|\right)$ is a normed space and T is an arbitrary subset of V ($T \subset V$), we say $u \in T$ is the best approximation of V members, for every member such as v of V we have:

$$\|u - v\| \to 0$$

In other words, $\tilde{u} \in T$ is the best approximation of the V members if for every $u \in T$, we have:

$$\|v - \tilde{u}\| \le \|v - u\|, \quad v \in V$$

4.5.6 Example

Suppose that $V = \mathbb{R}^2$ and $\|.\| = \|.\|_2$, that is, $\left(\mathbb{R}^2,\|.\|_2\right)$ and also $T = \left\{x \in V \,|\, \|x\| \le 1\right\}$. So V is the whole space of $\mathbb{R}^2$ with the norm of L_2 and T points on a circle with a radius of one. Obviously, a point like $\tilde{x}$ out of T can be found that for every u of V and every x of T:

$$\|\tilde{x} - u\| \le \|x - u\|$$

4.5.7 Example

Suppose that $T = \left\{u \in V \,|\, u(x) = e^{\beta x}, \beta > 0\right\}$ is a subset of $\left(c[0,1],\|.\|_\infty\right)$ and v is a constant function $v(x) = \dfrac{1}{2}$. We want to get the best approximation of $\tilde{u} \in T$.

The goal is to find a function like $\tilde{u}(x) = e^{\tilde{\beta} x}$, so that the maximum value of the following functions is minimized, that is,

$$\min_{\beta>0} \max_{0\le x\le 1} \left| \frac{1}{2} - e^{\bar{\beta}x} \right|$$

It is obvious that

$$\max_{0\le x\le 1} \left| \frac{1}{2} - e^{\bar{\beta}x} \right| = e^{\beta} - \frac{1}{2}$$

So

$$\inf_{\beta>0} \left(e^{\beta} - \frac{1}{2} \right) = \frac{1}{2}$$

Therefore, it can be claimed that the problem of finding best approximation has no answer. Why?

By providing these examples, we will express the best approximation.

4.5.8 Definition The Best Approximation

Suppose that T is a subset of a normed linear space of $(V, \|.\|)$ and $v \in V$. $\tilde{u} \in T$ is called the best approximation of v, provided that

$$\|v - \tilde{u}\| = \inf_{u\in T} \|v - u\|$$

If we assume

$$E_T(v) = \inf_{u\in T} \|v - u\|$$

then $E_T(v)$ is called the deviation of member v from set T. Obviously, in a special case, one can always assume $\tilde{u} = v$ and therefore $\|v - \tilde{u}\| = 0$.

4.5.9 Existence of the Best Approximation

The key difference between example (4.5.6) and example (4.5.7) is that set T on example (4.5.6) is a compact set of V while that in example (4.5.7) is not. We will now examine these two differences further.

4.5.10 Minimum Sequence

Sequence $(u_i)_{i\in\mathbb{N}}$ of the members of set $V \supset T$ is called a minimum sequence if for every $v \in V$, we have:

$$\lim_{i\to\infty} \|v - u_i\| = E_T(v)$$

Given the definition of the distance of $E_T(v)$, it is obvious that there is always a minimum sequence for each non-empty subset T and each member $v \in V$. But for a minimum sequence, the norm $\|v - u_i\|$ is required to be converged. Of course, for arbitrary subsets such as T, it cannot be claimed that this convergence is a sufficient condition for the convergence of (u_i) to a member of T or even a member of V. For this purpose, we present the following lemma.

Remark: If a group of a sequence converges to a point like u^*, then u^* is called a cluster point.

4.5.11 Lemma

Suppose that $v \in V$ and the cluster point u^* form a minimum sequence. If $u^* \in T$, then it is the best approximation of v out of T.

Proof: Suppose that (u_i) is a minimum sequence that

$$\lim_{i \to \infty} \|v - u_i\| = E_T(v)$$

Also assume that subsequence $\left(u_{j(i)} \right)$ converges to $u^* \in T$. In this case, given that

$$\lim_{i \to \infty} \|v - u_i\| = E_T(v), \qquad \lim_{j \to \infty} \|u_i - u^*\| = 0$$

we can say that for every j, we have:

$$\|u - u^*\| \leq \|v - u_i\| + \|u_j - u^*\|, \qquad \|v - u^*\| \leq E_T(v)$$

For every $u \in T$, we have

$$E_T(v) \leq \|v - u\|$$

So, it can be concluded that

$$\|v - u^*\| = E_T(v)$$

and u^* is the best approximation.

4.5.12 Theorem

Suppose that $T \subset V$ is a compact set. In this case, for every $v \in V$, a best approximation of $\tilde{u} \in T$ exists.

Proof: Suppose that $(u_i)_{i \in N}$, for every $v \in N$, is a minimum sequence. Since T is compact, then this minimum sequence contains a convergent subsequence. According to Lemma (4.5.11), this sub-sequence converges to the best approximation of $\tilde{u} \in T$.

4.5.13 Best Approximation Uniqueness

After discussing the existence of the best approximation, we move on to the discussion of the uniqueness of this approximation. If we pay attention to example (4.5.6), in this example, the best approximation is unique. Now the example can be changed in another way. Namely, suppose

$$\hat{T} = T/T^*, \quad T^* = \left\{ x \in V \middle| \|x\| \le 1, \quad x_1 0, x_2 > 0 \right\}$$

Obviously, both points of $(0,1)$ and $(1,0)$ are the best approximations of point $(1,1)$ of $\hat{T}$.

One of the conditions satisfied in example (4.5.6) was convexity of T. So, it seems that the convexity of T can be used as a precondition for the uniqueness of best approximation (Figure 4.1).

4.5.14 Definition Convex Set

Set T is called convex if

$$\forall u_1 \forall u_2 \forall \lambda \left((u_1, u_2 \in T \ \& \ u_1 \ne u_2 \ \& \ 0 < \lambda < 1) \Rightarrow \lambda u_1 + (1 - \lambda) u_2 \in T \right)$$

That is, for every two distinct arbitrary points, their connecting line segment is also completely inside T. Now if points $u_1 \ne u_2$ are all inner or internal points, we call set T strong convex. Strong convexity means that the boundaries of T do not contain any straight lines.

4.5.15 Theorem Uniqueness

Suppose that T is a strong compact and convex subset of a normed linear space V. In this case, for every $v \in V$, there is exactly one best approximation for v of T.
   Proof: Suppose that $\tilde{u}_1 \ne \tilde{u}_2$ are two approximations of $v \in V$. In this case,

$$\left\| \frac{1}{2}(\tilde{u}_1 + \tilde{u}_2) - v \right\| \le \frac{1}{2}\|\tilde{u}_1 - v\| + \frac{1}{2}\|\tilde{u}_2 - v\|$$

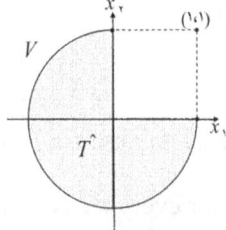

FIGURE 4.1
Best approximations of point $(1,1)$ of $\hat{T}$.

Then,

$$\left\|\frac{1}{2}(\tilde{u}_1 + \tilde{u}_2) - v\right\| \le E_T(v)$$

So

$$\left\|\frac{1}{2}(\tilde{u}_1 + \tilde{u}_2) - v\right\| = E_T(v)$$

And given that T is strong convex, then there exists $\lambda \in (0,1)$ such that

$$\tilde{u} := \frac{1}{2}(\tilde{u}_1 + \tilde{u}_2) + \lambda\left(v - \frac{1}{2}(\tilde{u}_1 + \tilde{u}_2)\right) \in T$$

If $\hat{\lambda} > 0$ is one of these λs, then:

$$\|\tilde{u} - v\| = \left\|\frac{1}{2}(1 - \hat{\lambda})(\tilde{u}_1 + \tilde{u}_2) - (1 - \hat{\lambda})v\right\| = (1 - \hat{\lambda})E_T(v)$$

Therefore, $\|\tilde{u} - v\| \le E_T(v)$ and this is a contradiction, so the induction assumption is false and the sentence is proved.

4.5.16 Theorem-Best Approximation Theory in the Normed Linear Space

Assume that U is a finite linear subspace of the normed linear space V. In this case, for every $v \in V$, there is at least one best approximation of $\tilde{u} \in U$.

   Proof: First, we show that each minimum sequence V is bounded. Suppose that $(u_i)_{i\in\mathbb{N}}$ is a minimum sequence in V corresponded to $v \in V$. Then,

$$\forall i \ge N, \quad E_V(v) \le \|v - u_i\| \le E_V(v) + 1$$

So, for $i \ge N$, we have:

$$\|u_i\| \le \|v - u_i\| + \|v\| \le E_T(v) + 1 + \|v\| := k_1$$

And also suppose that for every $i < N$, $k_2 \ge \|u_i\|$. Assuming $k := \max\{k_1, k_2\}$, we have:

$$\forall i \in \mathbb{N}, \quad \|u_i\| \le k$$

Now, since V is closed, the cluster point of u^* must belong to V, and according to Lemma (4.5.11), u^* is a best approximation of V.

4.5.17 Best Approximation in Spline Space

We observed that $S_l(\Omega_n)$ is a vector space with finite dimension of $n+1$. It can be shown that the polynomial $g \in P_{n-1}[x]$ is the best approximation for $f \in c[a,b]$ if and only if there is an ordered partition as $a \leq x_0 < \cdots < x_n \leq b$ such that for $i = 0,\ldots,n$

$$\max_{x_i}|f(x_i) - g(x_i)| = \|f - g\|_\infty$$

is minimized.

4.5.18 Definition

Point $\xi \in [x_i, x_{i+1})$ with $i = 0,1,\ldots,n-1$ is called a fundamental zero of the spline function $S \in S_l(\Omega_n)$ if $S(\xi) = 0$ and $S(x)$ are not zero at any other point of this interval. If $S(b) = 0$ (last point), then b is also a fundamental zero.

4.5.19 Example

Suppose that for $i = 0,\ldots,n-1$, $x_1 \in [x_i, x_{i+1})$ and $q_{l1} = (x - x_1)^l$.

$$q_{l1}(x_1) = 0$$

No other point is the zero value of this function on the above interval. Therefore, x_1 is a fundamental zero of order l (q_{l1}).

4.5.20 Example

Every $S(x)$ with $S \in S_l(\Omega_n)$ has at most n fundamental zeros of order l.

4.5.21 Theorem

Each spline function $S \in S_l(\Omega_n)$ has at most $n+l-1$ fundamental zeros on the interval $[a,b]$, which is considered as the repeat order of each fundamental zero.

Proof: Suppose that function $S(x)$ on the interval $[a,b]$ has r fundamental zeros. According to the Rolle's theorem, S' has $r - l$ fundamental zeros and S'' has $r - 2$ fundamental zeros, and finally $S^{(l-1)}$ has $r - l + 1$ fundamental zeros in $[a,b]$ and according to $S^{(l-1)} \in S_1(\Omega_n)$, therefore $S^{(l-1)}$ has at most n fundamental zero, that is,

$$r - l + 1 \leq n \implies r \leq n + l - 1$$

4.5.22 Lemma

If the spline function $S \in S_l(\Omega_n)$ for $x \in [x_0, x_v]$ and $x \in [x_\tau, x_n]$ is equal to zero, so that $\tau - v \geq l + 1$ and $0 < v < \tau < n$, but it is not zero anywhere else, then r fundamental zeros of S in (x_v, x_τ) are satisfied the following boundary:

$$r \leq \tau - (v + l + 1)$$

Proof: We define

$$\Omega_{\tau-v} = \{x_v, \ldots, x_\tau\}$$

So, the maximum fundamental zeros for each spline in space $S_l(\Omega_{\tau-v})$ is satisfied the bound of $r \leq \tau - v + l - 1$. Given that $S \in c^{l-1}[x_0, x_n]$, we have:

$$S(x_v) = S'(x_v) = \cdots = S^{(l-1)}(x_v) = 0$$
$$S(x_\tau) = S'(x_\tau) = \cdots = S^{(l-1)}(x_\tau) = 0$$

Therefore, x_τ and x_v are repeated roots and of the repeat order of l, $S(x) = 0$.

$$r \leq \tau - v + l - 1 - 2l = \tau - (v + l + 1)$$

And since $\Omega_{\tau-v} \subset \Omega_n$, then the sentence is true for every $S \in S_l(\Omega_n)$.

ı

4.5.23 Haar Condition

Assume that $\{h_i(x)\}_{i=1}^n$ is such that for every distinct point $x_1 < \cdots < x_n$, $\left| h_i(x_j) \right|_{i,j=1}^n$ is not zero for every m, $m \leq n$, then $\{h_i(x)\}_{i=1}^n$ satisfy Haar condition on the interval $[x_1, x_n]$.

4.5.24 Remark

Suppose that $\{x_i\}_{i=1}^n$ are n distinct points over $[a,b]$ such that $\{h_i(x)\}_{i=1}^n$ satisfy Haar condition over $[a,b]$, then the following system of equations for $j = 1, \ldots, n$ has a unique answer.

$$\sum_{i=1}^n w_i h_i(x_j) = f_i$$

4.5.25 Haar Space

Suppose that $g_1, \ldots, g_n \in c[a,b]$ have n independent linear functions such that every g, $g \in span\{g_i\}_{i=1}^n$ has (at most) $n - 1$ roots. In this case, $V = span\{g_i\}_{i=1}^n$ is called a Haar space.

4.5.26 Example

Suppose

$$V = span\{g_i\}_{i=1}^n = \left\{ \sum_{i=1}^n a_i x^{i-1} \middle| g_i(x) = x^{i-1}, \quad i = 1,...,n \right\}$$

In this case, since each point g of space V is a polynomial of at most degree $n-1$, then, it has at most $n-1$ roots, so V is a Haar space.

Question: Is space $S_l(\Omega_n)$ a Haar space? Why?

4.5.27 Remark

If $V = span\{g_i\}_{i=1}^n$ is a Haar space of $c[a,b]$, then for every $f \in c[a,b]$, there is the unique best approximation such as $\tilde{f} \in V$ so that:

$$\forall f \left(\bar{f} \in V \Rightarrow \|f - \tilde{f}\| \le \|f - \bar{f}\| \right)$$

4.5.28 Types of Splines

In this section, the types of splines in vector space $S_l(\Omega_n)$ for $l = 2m-1$ and $m \ge 2$ are discussed.

It is obvious that

$$\dim(S_l(\Omega_n)) = n + 2m - 1$$

Given that

$$\forall S (S \in S_l(\Omega_n) \Rightarrow S(x_i) = f_i, \quad i = 0,...,n)$$

To obtain this number of equations, we have the following conditions:

1. Suppose that $f \in c^m[a,b]$ and $2 \le m \le n+1$ and

$$S^{(\mu)}(a) = S^{(\mu)}(b) = 0, \quad \mu = m,...,2m-2$$

In this case, a natural spline is obtained.

2. Assuming $f \in c^m[a,b]$ and $2 \le m \le n+1$ and

$$S^{(\mu)}(a) = f^{(\mu)}(a), \quad S^{(\mu)}(b) = f^{(\mu)}(b), \quad \mu = 1,...,m-1$$

we have a bounded (Hermite) spline.

3. Assuming $f \in c^m[a,b]$ and

$$f^{(k)}(a) = f^{(k)}(b), \quad k = 0,1,\ldots,m-1$$
$$S^{(\mu)}(a) = S^{(\mu)}(b), \quad \mu = 1,\ldots,2m-2$$

we have periodic splines.

4.5.29 Remark-Integral Relation

If $f \in c^m[a,b]$ and

$$S \in S_{2m-1}(\Omega_n), \quad m \geq 2$$

is the spline interpolating function, so that $d(x) = f(x) - S(x)$ and satisfies the following boundary conditions

$$\sum_{\mu=0}^{m-2} (-1)^\mu S^{(m+\mu)}(a) d^{(m-\mu-1)}(a) = \sum_{\mu=0}^{m-2} (-1)^\mu S^{(m+\mu)}(b) d^{(m-\mu-1)}(b)$$

Then,

$$\int_a^b \left(f^{(m)}(x) \right)^2 dx = \int_a^b \left(f^{(m)}(x) - S^{(m)}(x) \right)^2 dx + \int_a^b \left(S^{(m)}(x) \right)^2 dx$$

4.5.30 Remark

The problem of natural, bounded, periodic spline interpolation always has a unique answer.

4.5.31 Remark

Suppose that $f \in c^m[a,b]$, $m \geq 2$, and $S \in S_{2m-1}(\Omega_n)$ is the spline interpolating function of the above types, in this case:

$$\left\| f^{(j)} - S^{(j)} \right\|_\infty \leq \frac{m!}{\sqrt{m}} \cdot \frac{1}{j!} \cdot h^{m-j-\frac{1}{2}} \cdot \left\| f^{(m)} \right\|_2, \quad j = 0,1,\ldots,m-1$$

where $h = \max\limits_{0 \leq i \leq n-1} |x_{i+1} - x_i|$.

The application of theorem (4.5.31), for $m = 2$ (cubic) is as follows:

$$j = 0 \Rightarrow \|f - S\|_\infty \leq \sqrt{2} \cdot h^{\frac{3}{2}} \cdot \|f''\|_2$$

$$j = 1 \Rightarrow \|f' - S'\|_\infty \leq \sqrt{2} \cdot h^{\frac{1}{2}} \cdot \|f''\|_2$$

According to theorem (4.5.31), it can be claimed that spline convergence is of uniform type. That is, there is convergence across points within the definition domain. For further explanation, we approximate function $f(x) = \dfrac{1}{1+x^2}$ on the interval $[-5,5]$ with a cubic spline. It is observed that the more the number of domain partition points, the closer and more convergent the approximation is to the function f, so that with $n = 19$, almost the function f itself is plotted (Figures 4.2–4.5).

4.5.32 B-Spline

In this section, we introduce another basis for the space $S_l\left(\Omega_n\right)$ other than the spline basis, which is useful for calculations with the splines.

This basis was the basic spline curves, which was later renamed B-spline. In this topic, we work with spline spaces of infinite dimensions and show the existence of definite elements with compact points that can be used as basic functions.

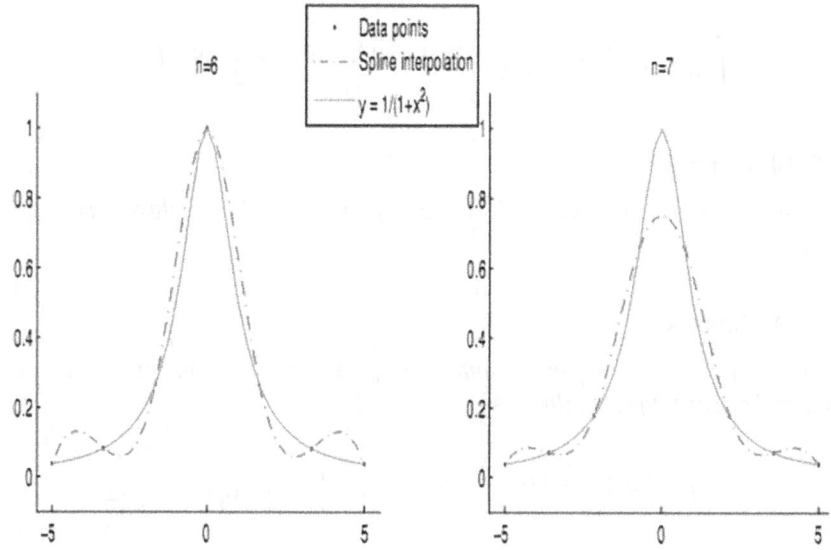

FIGURE 4.2
Convergence with $n = 6, 7$.

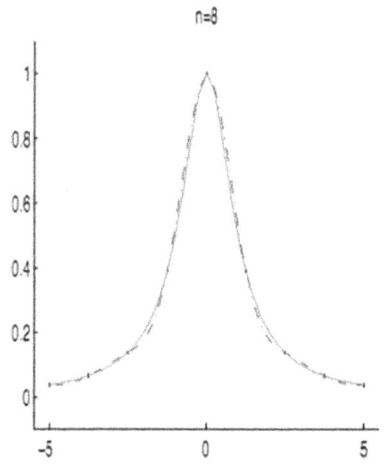
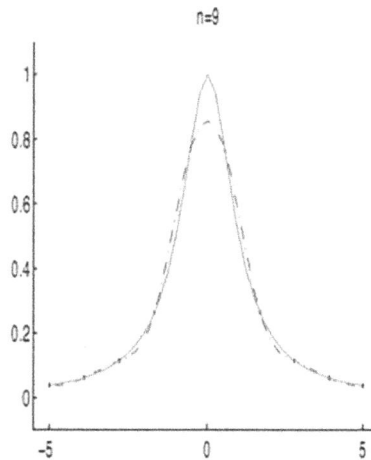

FIGURE 4.3
Convergence with $n = 8, 9$.

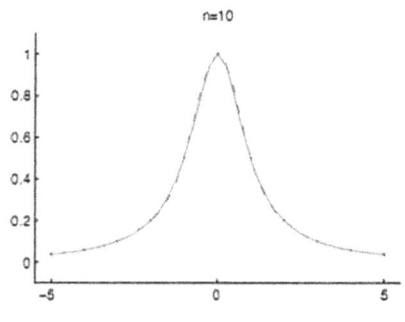
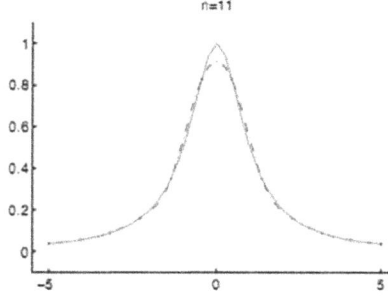
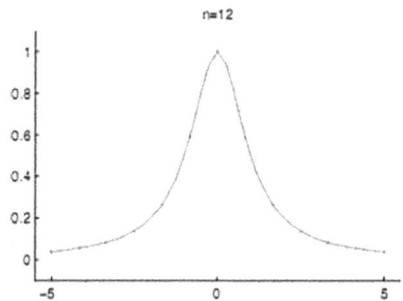
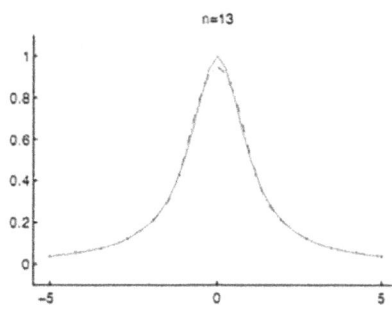

FIGURE 4.4
Convergence with $n = 10 - 12$.

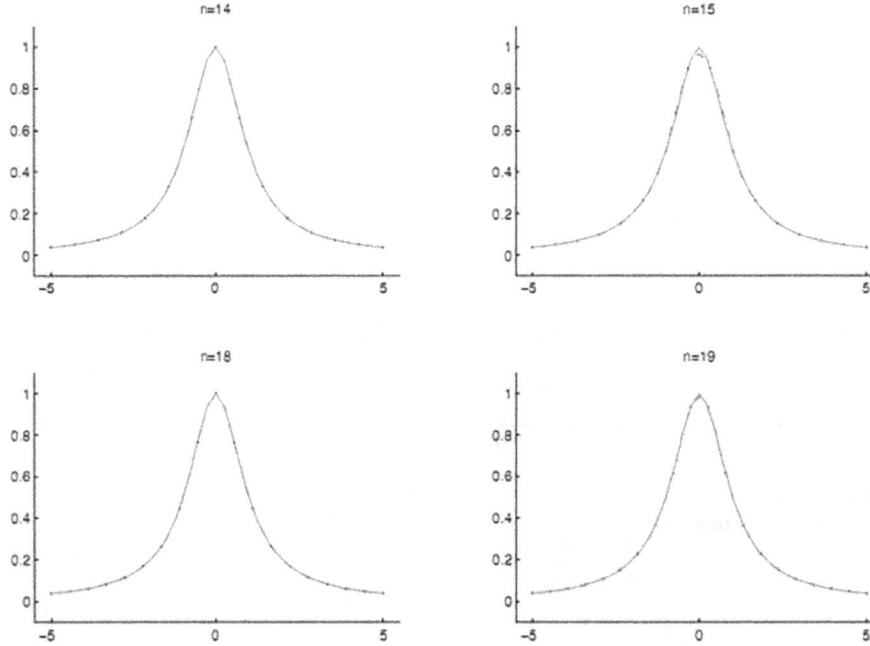

FIGURE 4.5
Convergence with $n = 14 - 1$.

4.5.33 Existence of B-Spline

The infinite number of points $\Omega_\infty = \{x_i\}_{i \in \mathbb{Z}}$ with $x_i < x_{i+1}$ are considered so that

$$
\begin{cases}
i \to -\infty \Rightarrow x_i \to -\infty \\
i \to \infty \;\; \Rightarrow x_i \to \infty
\end{cases}
$$

We have to show that for every $i \in \mathbb{Z}$, there is exactly one spline $S \in S_l(\Omega_\infty)$, so that $S(x) = 0$ for $x_{i+l+1} \geq x$ and $x < x_i$, and if the normalization condition is also met, then we have:

$$
\int_{-\infty}^{\infty} S(x)dx = \int_{x_i}^{x_{i+l+1}} S(x)dx = 1
$$

Proof: For $x \in [x_{i-1}, x_{i+l+2}]$, it is clear that the expansion of spline S in terms of one-way bases as

$$
S(x) = \sum_{i=0}^{l} a_i x_i + \sum_{i=1}^{n-1} b_i (x - x_i)_+^l
$$

does not include all or part of the basis polynomials because for $x < x_i$, $S(x) = 0$. So, we can say that on the above-mentioned interval:

$$S(x) = \sum_{k=0}^{m} b_k (x - x_{i+k})_+^l$$

which is determined for some values of m. Now if we assume $m = l + 1$, we have:

$$S(x) = 0, \quad x \geq x_{i+l+1}$$

and also,

$$(x - x_{i+k})_+^l = (x - x_{i+k})^l, \quad k = 0, 1, \ldots, m$$

Therefore, it can be concluded that the coefficients $b_0, \ldots, b_{l+1}$ are the unique answers of the following non-singular system of equations:

$$\sum_{k=0}^{l+1} b_k (x - x_{i+k})_+^l = 0, \quad x \geq x_{i+l+1}$$

$$\begin{cases} b_0 + b_1 + \cdots + b_{l+1} = 0 \\ b_0 x_i + b_1 x_{i+1} + \cdots + b_{l+1} x_{i+l+1} = 0 \\ \vdots \\ b_0 x_i^l + b_1 x_{i+1}^l + \cdots + b_{l+1} x_{i+l+1}^l = 0 \end{cases}$$

Now we add a normalization condition

$$\sum_{k=0}^{l+1} \frac{b_k}{l+1} (x_{i+l+1} - x_{i+k})^{l+1} = 1$$

That is,

$$b_0 x_i^{l+1} + b_1 x_{i+1}^{l+1} + \cdots + b_{l+1} x_{i+l+1}^{l+1} = (-1)^{l+1} (l+1)$$

So, a system is created with $l+2$ equations and $l+2$ unknowns where the unknowns are $b_0, \ldots, b_{l+1}$, and the determinants of the system is Vandermonde, which is the non-zero; therefore, the system has a trivial unique answer.

Note: If $m \leq l$, then the conditions $S(x) = 0$ for $x < x_i$ and $S(x) = 0$ for $x \geq x_{i+k}$ result in $S(x) = 0$. In fact, in this case, the above-mentioned system of equations has at least one column less, so that $l+1$ homogenized equations are remained for $m + 1 \leq l + 1$ unknown of $b_0, \ldots, b_m$, and the intended matrix is

Vandermonde, and the system has a unique zero answer as $b_0 = \cdots = b_m = 0$. It can be concluded that there is not any nontrivial spline of degree l that over a subset of supporting points is a B-spline with the same degree of snake, that is, B-spline has minimal supporting points.

4.5.34 Definition

B-spline of degree l, corresponding to nodes t_v in the set of points Ω_∞ is defined as follows:

$$B_{lv}(x) = (t_{v+l-1} - t_v) q_l(\cdot, x)[t_v, \ldots, t_{v+l+1}]$$

Spline can also be written as:

$$B_{lv}(x) = \sum_{k=v}^{v+l+1} \left(\prod_{r=v, r \neq k}^{v+l+1} (t_k - t_r)^{-1} \right) (t_k - x)_+^l$$

For this purpose, it is enough to check that B_{lv} interpolates all points on $[t_v, t_{v+l+1}]$, where B_{lv} is the B-spline. For this purpose, we consider points $x < t_v$. In this case, $q_l(\cdot, x) \in P_l$ and the divided differences of order $(l+1)$ is zero, namely:

$$q_l(\cdot, x)[t_v, \ldots, t_{v+l+1}] = 0$$

Otherwise, for $t_{v+l+1} \leq x$, we have:

$$(t - x)_+^l = q_l(t, x) = 0$$

We showed that for $x < t_v$ and $x \geq t_{v+l+1}$, $B_{lv}(x) = 0$, and for $x \in (t_v, t_{v+l+1})$, B_{lv} it is not zero, and we have positive B splines.

4.5.35 B-Spline Positivity

Spline B_{lv} has at most $2l$ zeros in $[t_v, t_{v+l+1}]$, where each of points t_v and t_{v+l+1} are repeated zeros of B_{lv} with the order of l, because $B_{lv} \in c^{l-1}(-\infty, \infty)$. It can be concluded that on the interval (t_v, t_{v+l+1}), no root can exist. On the interval (t_v, t_{v+l+1}), B_{lv} is not zero, so if it is negative, it must intersect the x axe at a point, in which case a zero is obtained; therefore, it is positive. In fact, on the interval (t_{v+l}, t_{v+l+1}), we have:

$$B_{lv}(x) = \left(\prod_{r=v}^{v+l} (t_{v+l+1} - t_r) \right)^{-1} (t_{v+l+1} - x)^l > 0$$

Now, consider the space $S_l(\Omega_n)$ of the l degree splines that are defined over $[x_0, x_n]$. In the defined B-spline set, only $B_{l,n-1}, \ldots, B_{l,-l}$ in $[x_0, x_n]$ have non-zero values. We show that these $n+l$ spline functions form a basis. So we prove that they are linearly independent.

To prove the linear independence of $B_{l,n-1}, \ldots, B_{l,-l}$, we must prove that if for every $x \in [x_0, x_n]$,

$$S(x) = \beta_{-l} B_{l,-l}(x) + \cdots + \beta_{n-1} B_{l,n-1}(x) = 0$$

Then, $\beta_{-l} = \cdots = \beta_{n-1} = 0$.

To do this, we add node x_{-l-1}, with $x_{-l-1} < x_{-l}$. Using the B-spline definition for $x \in [x_{-l-1}, x_{-l}]$, we have $S(x) = 0$. Now if $S(x) = 0$ is zero over $[x_0, x_1]$, then the number of fundamental zeros in $[x_{-l}, x_0]$ satisfies the following condition.

$$\tau + v = l < l + 1 \Rightarrow r \le \tau - (v + l + 1) < 0$$

Therefore, it can be concluded that $S(x) = 0$ over $[x_{-l}, x_0]$, so on the whole interval, we have $S(x) = 0$.

We have already shown that $S(x)$ being zero on the whole interval of $[x_0, x_n]$ is similar to $S(x)$ being zero on the interval $[x_{-l}, x_n]$. Therefore, for linear independence on the interval $[x_0, x_n]$, it is enough to examine the linear independence in $[x_{-l}, x_n]$.

Now, to check the linear independence of $B_{l,n-1}, \ldots, B_{l,-l}$ on the interval $[x_{-l}, x_n]$, we do this over the subintervals.

If $S(x) = 0$ over $[x_{-l}, x_{-l+1}]$, then $S(x) = \beta_{-l} B_{l,-l}(x) = 0$ where $B_{l,-l}(x) > 0$ so $\beta_{-l} = 0$. Now over $[x_{-l+1}, x_{-l+2}]$, if $S(x) = \beta_{-l+1} B_{l,-l+1} = 0$, then $\beta_{-l+1} = 0$. If we continue in this way, the coefficients will be zero on the whole interval and therefore it has linear independence.

So, we show that B-splines $B_{l,n-1}, \ldots, B_{l,-l}$ form a basis over $[x_0, x_n]$ for space $S_l(\Omega_n)$.

4.5.36 Theorem (Representation)

Every spline $S \in S_l(\Omega_n)$ over $[x_0, x_n]$ has a unique expansion in terms of B-splines as follows:

$$S = \sum_{v=-l}^{n-1} \alpha_v B_{lv}, \quad \alpha_v \in \mathbb{R}$$

4.5.37 Other Properties of B-Splines

B-spline B_{lv} forms a single partition as

$$\sum_{v \in Z} B_{lv}(x) = 1, \quad x \in (-\infty, \infty)$$

Proof: If $l = 0$, then, for every x of $[x_v, x_{v+1})$, $B_{0v}(x) = 1$ and $B_{0,v}(x) = 0$ on other intervals, then

$$\sum_{v \in Z} B_{0v}(x) = 1$$

Now for $l \leq 1$, we can write B_{lv} as follows:

$$B_{lv}(x) = q_l(\cdot, x)[t_{v+1}, \ldots, t_{v+l+1}] - q_l(\cdot, x)[t_v, \ldots, t_{v+l}]$$

So, we can conclude that for every x of $[t_\mu, t_{\mu+1}]$, we have:

$$\sum_{v \in Z} B_{lv}(x) = \sum_{v=\mu-1}^{\mu} B_{lv}(x)$$

$$= \sum_{v=\mu-l}^{\mu} q_l(\cdot, x)[t_{v+1}, \ldots, t_{v+l+1}] - \sum_{v=\mu-l}^{\mu} q_l(\cdot, x)[t_v, \ldots, t_{v+l}]$$

$$= q_l(\cdot, x)[t_\mu, \ldots, t_{\mu+l+1}] - q_l(\cdot, x)[t_{\mu-l}, \ldots, t_\mu]$$

The divided differences of the order l with respect to t from a polynomial $(t-x)^l$ is

$$q_l(\cdot, x)[t_{\mu+1}, \ldots, t_{\mu+l+1}] = 1$$

which is the coefficient of the leading sentence, so its value is one. While for $x \in [t_\mu, t\mu+1]$, $q_l(t, x) = (t, x)_+^l$ that

$$q_l(\cdot, x)[t_{\mu-l}, \ldots, t_\mu] = 0, \quad t_{\mu-l}, \ldots, t_\mu < x$$

Then,

$$\sum_{v \in Z} B_{lv}(x) = 1$$

And the recursive formula is as follows:

$$B_{lv}(x) = \frac{x - x_v}{x_{v+l} - x_v} B_{l-1,v}(x) + \frac{x_{v+l+1} - x_v}{x_{v+l+1} - x_{v+1}} B_{l-1,v-1}(x)$$

Now let us investigate other properties of spline.

Remark:

$$\overrightarrow{R(t)} = f(x)\vec{i} + g(t)\vec{j} + k(t)\vec{k}$$

$$k = \frac{\|v \times a\|}{\|v^3\|} \text{ curvature coefficient}$$

$$v = R'(t), \quad a = R''(t)$$

If $y = f(x)$, then we put

$$x = t, \quad y = f(t)$$

So, we will have

$$\overrightarrow{R(t)} = t\vec{i} + f(t)\vec{j}$$

$$v = R'(t) = \vec{i} + f'(t)\vec{j}$$

$$a = R''(t) = f''(t)\vec{j}$$

So, the curvature coefficient of the curve will be as follows:

$$k = \frac{\|f''(t) \cdot \vec{k}\|}{\|\vec{i} + f'(t)\vec{j}^3\|}$$

Now if $|f'(t)| < 1$, $f'^2(t)$ tends to zero, so we have $k \simeq f''(t)$.

4.5.38 Problem

Explain how the approximation curve of a function such as f is smoothed by the spline interpolating function.

Solution: Given that in natural spline, $s''(x_0) = s''(x_n) = 0$, so the curvature is zero, that is, the curve is smooth, and it can also be claimed that natural spline will be a better approximation than the bound and periodic spline.

4.5.39 Problem

A set of nodes such as $\Delta = \{x_0 < x_1 < \cdots < x_n\}$ and a set of values such as $Y = \{y_0, \ldots, y_n\}$ are given. Prove that the spline function $S_\Delta(y, .)$ with condition

$$S_\Delta''(y, x_0) = S_\Delta''(y, x_n) = 0$$

is unique.

Solution: Suppose that $f \in k^2[x_0, x_n]$ and $S_\Delta(y, \cdot)$ is the cubic spline interpolating function at points $x_0, \dots, x_n$, then

a. $S_\Delta \in c^2[x_0, x_n]$

b. On the each subinterval $[x_j, x_{j+1}]$, $j = 0, \dots, n-1$, the function S_Δ is a polynomial of at most degree 3.

c. $S_\Delta(y, x_i) = f(x_i) = y_i$.

d. The cube spline function S_Δ is a function of polynomial piecewise function of degree 3 in which the values of these polynomials and their first and second derivatives are equal in the middle nodes, that is,

$$S_\Delta^{(k)}(y, x_i^+) = S_\Delta^{(k)}(y, x_i^-) = 0, \quad k = 0, 1, 2, \quad i = 1, \dots, n-1$$

Suppose that $\bar{S}_\Delta(y, \cdot)$ is another spline function that has the same properties as $S_\Delta(y, \cdot)$. Prove that S_Δ and $\bar{S}_\Delta$ are equal.

It suffices to prove that in each of the subintervals $[x_j, x_{j+1}]$, $j = 0, \dots, n-1$, the spline functions S_Δ and $\bar{S}_\Delta$ are equal.

For this purpose, we consider the interval $[x_j, x_{j+1}]$, $j = 1, \dots, n-2$.

According to property (b), S_Δ'' and $\bar{S}_\Delta''$ in these subintervals are of at most first degree. We put

$$P_j(x) = S_\Delta''(y, x) = \bar{S}_\Delta''(y, x), \quad x \in [x_j, x_{j+1}]$$

where $P_j(x)$ is a polynomial of at most degree 1. On the other hand,

$$P_j(x_j) = S_\Delta''(y, x_j) = \bar{S}_\Delta''(y, x_j) = f_j'' - f_j'' = 0$$

$$P_j(x_{j+1}) = S_\Delta''(y, x_{j+1}) = \bar{S}_\Delta''(y, x_{j+1}) = f_{j+1}'' - f_{j+1}'' = 0$$

So $P_j(x)$ has two roots on every subinterval, which contradicts the fundamental theorem of algebra. Thus,

$$P_j(x) \equiv 0$$

So, it can be said that:

$$S_\Delta''(y, x) = \bar{S}_\Delta''(y, x), \quad x \in [x_j, x_{j+1}], \quad j = 1, \dots, n-2$$

Now, we have

$$P_0(x) = S_\Delta''(y, x) = \bar{S}_\Delta''(y, x)$$

on the subinterval $[x_0, x_1]$, where $P_0(x)$ is a polynomial of at most degree 1. On the other hand,

$$P_0(x_0) = S_\Delta''(y, x_0) = \bar{S}_\Delta''(y, x_0) = 0$$

$$P_0(x_1) = S_\Delta''(y, x_1) = \bar{S}_\Delta''(y, x_1) = 0$$

So, $P_0(x)$ has two roots on this interval and this contradicts the fundamental algebra theorem. Therefore:

$$P_0(x) \equiv 0$$

and, hence

$$S_\Delta''(y, x) = \bar{S}_\Delta''(y, x), \quad x \in [x_0, x_1]$$

Consequently, we have

$$S_\Delta''(y, x) = \bar{S}_\Delta''(y, x), \quad x \in [x_{n-1}, x_n]$$

As a result of the above-mentioned equations:

$$S_\Delta''(y, x) = \bar{S}_\Delta''(y, x), \quad c \in [x_j, x_{j+1}], \quad j = 0, \ldots, n-1$$

Then,

$$S_\Delta'(y, x) = \bar{S}_\Delta'(y, x) + a$$

As $S_\Delta'(y, x_j) = f_j'$, $j = 1, \ldots, n-1$ and according to the property (d), $a = 0$, that is,

$$S_\Delta'(y, x) = \bar{S}_\Delta'(y, x)$$

Therefore,

$$S_\Delta(y, x) = \bar{S}_\Delta(y, x) + b$$

According to the property (d), it can be concluded that $b = 0$. That is, on every subinterval, $S_\Delta(y, x) = \bar{S}_\Delta(y, x)$ and the natural spline is unique.

4.5.40 Problem

Consider the partition $\Delta := \{a = x_0 < x_1 < \cdots < x_n = b\}$ of the interval $[a,b]$ and the set of values $Y = \{y_0,\ldots,y_n\}$ and the function f with the conditions $f \in k^2(a,b)$, $f(x_i) = y_i$, $i = 0,\ldots,n$. Also let Δ' be a partition of the interval $[a,b]$ so that $\Delta \subset \Delta'$ and Y' are set of values for the points of the partition Δ'. Show that for any of (1), (2) or (3) cases, we have:

$$\|f\| \geq \|S_{\Delta'}(y',.)\| \geq \|S_\Delta(y,.)\|$$

1. $S''_\Delta(y,a) = S''_\Delta(y,b) = 0$
2. $f \in k_p^2[a,b]$,   periodic $S_\Delta(y,.)$
3. $S'_\Delta(y,a) = f'(a)$, $S'_\Delta(y,b) = f'(b)$

Solution: First we prove that:

$$\|f\| \geq \|S_\Delta(y,.)\|$$

$$\|f - S_\Delta\|^2 = \int_a^b |f''(x) - S''_\Delta(x)|^2\, dx$$

$$= \|f\|^2 - 2\int_a^b f''(x) \cdot S''_\Delta(x)\,dx + \|S_\Delta\|^2$$

$$= \|f\|^2 - 2\int_a^b \big(f''(x) - S''_\Delta(x)\big) \cdot S''_\Delta(x)\,dx - \|S_\Delta\|^2$$

As we know

$$\int_a^b \big(f''(x) - S''_\Delta(x)\big) \cdot S''_\Delta(x)\,dx = \sum_{i=1}^n \int_{x_{i-1}}^{x_i} \big(f''(x) - S''_\Delta(x)\big) S''_\Delta(x)\,dx$$

So, we calculate $\displaystyle\sum_{i=1}^n \int_{x_{i-1}}^{x_i} \big(f''(x) - S''_\Delta(x)\big) S''_\Delta(x)\,dx$

If we apply the integration by parts twice, we have:

$$\int_{x_{i-1}}^{x_i} \big(f''(x) - S''_\Delta(x)\big) S''_\Delta(x)\,dx = \Big[\big(f'(x) - S'_\Delta(x)\big) S''_\Delta(x)\Big]_{x_{i-1}^+}^{x_i^-}$$

$$- \Big[\big(f(x) - S_\Delta(x)\big) S'''_\Delta(x)\Big]_{x_{i-1}^+}^{\mu_i^-}$$

$$+ \int_{x_{i-1}}^{x_i} \big(f(x) - S_\Delta(x)\big) S_\Delta^{(4)}(x)\,dx$$

Because the spline function is of at most degree 3, so $S_\Delta^{(4)}(x) \equiv 0$ and as $f'(x), S_\Delta'(x)$ and $S_\Delta''(x)$ are continuous on the interval $[a,b]$, so we have:

$$\sum_{i=1}^{n} \int_{x_{i-1}}^{x_i} \left(f''(x) - S_\Delta''(x)\right) S_\Delta''(x) dx = \sum_{i=1}^{n} \left(\left[\left(f'(x) - S_\Delta'(x)\right) S_\Delta''(x) \right]_{x_{i-1}^+}^{x_i^-} \right)$$

$$- \sum_{i=1}^{n} \left(\left[\left(f(x) - S_\Delta(x)\right) S_\Delta'''(x) \right]_{x_{i-1}^+}^{x_i^-} \right)$$

$$= \left[\left(f'(x) - S_\Delta'(x)\right) S_\Delta''(x) \right]_a^b$$

$$- \sum_{i=1}^{n} \left(\left[\left(f(x) - S_\Delta(x)\right) S_\Delta'''(x) \right]_{x_{i-1}^+}^{x_i^-} \right)$$

Therefore, according to the above relations, we will have:

$$\|f - S_\Delta\|^2 = \|f\|^2 - 2\left[\left(f'(x) - S_\Delta'(x)\right) S_\Delta''(x) \right]_a^b$$

$$+ 2 \sum_{i=1}^{n} \left(\left[\left(f(x) - S_\Delta(x)\right) S_\Delta'''(x) \right]_{x_{i-1}^+}^{x_i^-} \right) - \|S_\Delta\|^2$$

According to the interpolation condition, the term $\displaystyle\sum_{i=1}^{n} \left(\left[\left(f(x) - S_\Delta(x)\right) S_\Delta'''(x) \right]_{x_{i-1}^+}^{x_i^-} \right)$ is always equal to zero. The term $\left[\left(f'(x) - S_\Delta'(x)\right) S_\Delta''(x) \right]_a^b$ is also equal to zero according to the cases (1), (2), and (3), then

$$0 \le \|f - S_\Delta\|^2 = \|f\|^2 - \|S_\Delta\|^2$$

Hence,

$$\|f\| \ge \|S_\Delta(y,.)\|$$

Now, we prove that S_Δ is unique.

By contradiction, we assume that there are at least two spline functions S_Δ and $\bar{S}_\Delta$ that are unique and satisfy the mentioned conditions. As

$$\|f\|^2 \ge \|S_\Delta\|^2 \Rightarrow \int_a^b |f''(x)|^2 dx \ge \int_a^b |S_\Delta''(x)|^2 dx$$

and according to the definition of $k^2[a,b]$, we have:

$$f'' \in L^2[a,b] \Rightarrow \int_a^b |f''|^2 < \infty$$

So,

$$\int_a^b |S_\Delta''|^2 < \infty$$

As S_Δ is real and $S_\Delta' \in c[a,b]$, therefore,

$$S_\Delta \in k^2[a,b]$$

And also,

$$\bar{S}_\Delta \in k^2[a,b]$$

Hence, f can be replaced by $\bar{S}_\Delta$:

$$\left\| \bar{S}_\Delta(x) - S_\Delta(x) \right\|^2 = \left\| \bar{S}_\Delta(x) \right\|^2 - \left\| S_\Delta(x) \right\|^2 \geq 0$$

So

$$\left\| \bar{S}_\Delta(x) \right\|^2 \geq \left\| S_\Delta(x) \right\|^2 \tag{4.14}$$

f can be replaced by S_Δ, too:

$$\left\| S_\Delta(x) \| - \| \bar{S}_\Delta(x) \right\|^2 = \left\| S_\Delta(x) \right\|^2 - \left\| \bar{S}_\Delta(x) \right\|^2 \geq 0$$

Then,

$$\left\| S_\Delta(x) \right\|^2 \geq \left\| \bar{S}_\Delta(x) \right\|^2 \tag{4.15}$$

According to the equations (4.14) and (4.15):

$$\left\| S_\Delta(x) \right\|^2 = \left\| \bar{S}_\Delta(x) \right\|^2$$

Then,

$$\left\| S_\Delta(x) - \bar{S}_\Delta(x) \right\|^2 = 0$$

According to the definition of norm

$$\int_a^b \left| S_\Delta''(y,x) - \bar{S}_\Delta''(y,x) \right| dx = 0$$

So,

$$\left| S_\Delta''(y,x) - \bar{S}_\Delta''(y,x) \right| = 0$$

Therefore,

$$S_\Delta''(y,x) - \bar{S}_\Delta''(y,x) = 0$$

Using integration, we have:

$$\int_a^b \left(S_\Delta''(x) - \bar{S}_\Delta''(x) \right) dx = 0$$

So,

$$\int_a^b S_\Delta''(x) dx - \int_a^b \bar{S}_\Delta''(x) dx = 0$$

Therefore,

$$S_\Delta(x) - \bar{S}_\Delta(x) + d_1 x + d_2 = 0$$

According to the interpolation condition in points a and b, we can write:

$$S_\Delta(a) - \bar{S}_\Delta(a) + d_1 a + d_2 = 0$$
$$S_\Delta(b) - \bar{S}_\Delta(b) + d_1 b + d_2 = 0$$

And therefore, $d_1 a + d_2 = 0$ and $d_1 b + d_2 = 0$ that requires $d_1 = d_2 = 0$ and so,

$$S_\Delta(x) = \bar{S}_\Delta(x)$$

In accordance, we can prove that on the partition of Δ':

$$\|f\| \geq \|S_{\Delta'}\|$$

And also, $S_{\Delta'}$ is a unique spline function. Now, we prove that: $\|S_{\Delta'}\| \geq \|S_\Delta\|$.

Since, the spline function S_Δ is unique on the partition Δ, then $S_{\Delta'}$ is not a spline interpolating function on the partition Δ and given that $S_{\Delta'} \in k^3(a,b)$, so $S_{\Delta'} \in k^2(a,b)$ and therefore, $S_{\Delta'}$ can be considered as an interpolating function like f so that:

$$\left\| S_{\Delta'}(x) - S_\Delta(x) \right\|^2 = \left\| S_{\Delta'}(x) \right\|^2 - \left\| S_\Delta(x) \right\|^2 \geq 0$$

Then,

$$\left\| S_{\Delta'}(x) \right\| \geq \left\| S_\Delta(x) \right\|$$

4.5.41 Problem

Suppose that $f \in k^4(a,b)$ and $S_\Delta(x)$ are spline interpolating functions that interpolate node points $\Delta = \{a = x_0 < x_1 < \cdots < x_n = b\}$. Show that:

$$\|f - S_\Delta\|^2 = \int_a^b \left(f(x) - S_\Delta(x)\right) f^{(4)}(x) dx$$

if only one of the following conditions is satisfied:

1. $f'(x) = S'_\Delta(x)$ $x = a,b$
2. $f''(x) = S''_\Delta(x)$ $x = a,b$
3. $f \in k_p^4(a,b)$,   periodic S_Δ

Solution:

$$\|f - S_\Delta\|^2 = \int_a^b \left|f''(x) - S''_\Delta(x)\right|^2 dx$$

$$= \int_a^b \left(f''(x) - S''_\Delta(x)\right)\left(f''(x) - S''_\Delta(x)\right) dx$$

$$= \int_a^b \left(f''(x) - S''_\Delta(x)\right) f''(x) dx - \int_a^b \left(f''(x) - S''_\Delta(x)\right) S''_\Delta(x) dx$$

$$:= I$$

If we apply integration by parts twice on each of above integrals, we will have:

$$I = \left[\left(f'(x) - S'_\Delta(x)\right) f''(x)\right]_a^b - \left[\left(f(x) - S_\Delta(x)\right) f'''(x)\right]_a^b$$

$$+ \int_a^b \left(f(x) - S_\Delta(x)\right) f^{(4)}(x) dx - \left[\left(f'(x) - S'_\Delta(x)\right) S''_\Delta(x)\right]_a^b$$

$$- \left[\left(f(x) - S_\Delta(x)\right) S'''_\Delta(x)\right]_{x_{i-1}^+}^{x_i^-} - \sum_{i=1}^n \int_{x_{i-1}}^{x_i} \left(f(x) - S_\Delta(x)\right) S_\delta^{(4)}(x) dx$$

$$= \left[\left(f'(x) - S'_\Delta(x)\right)\left(f''(x) - S''_\Delta(x)\right)\right]_a^b - \left[\left(f(x) - S_\Delta(x)\right) f'''(x)\right]_u^b$$

$$+ \int_a^b \left(f(x) - S_\Delta(x)\right) f^{(4)}(x) dx + \left[\sum_{i=1}^n \left(f(x) - S_\Delta(x)\right) S'''_\Delta(x)\right]_{x_{i-1}^+}^{x_i^-}$$

According to the interpolation condition, the second and forth terms are equal to zero. Considering conditions (1) and (2), the first term become zero. If condition (3) is met, then we have:

$$S'_\Delta(a) = S'_\Delta(b), \quad S''_\Delta(a) = S''_\Delta(b)$$
$$f'(a) = f'(b), \quad f''(a) = f''(b)$$

and therefore, the first term is equal to zero.

4.5.42 Problem

We define spline functions S_j in equidistant nodes of $x_i = a + ih$, $i = 0,...,n$ and $h > 0$ as follows:

$$S_j(x_k) = \begin{cases} 1 & j = k \\ 0 & j \neq k \end{cases}$$

where $j, k = 0,...,n$ and

$$S''_j(x_0) = S''_j(x_n) = 0$$

Check that the torques $M_1,...,M_{n-1}$ of S_j are as follows:

$$
\begin{cases}
M_i = -\dfrac{1}{\rho_i} M_{i+1}, & i = 1,...,j-2 \\[2ex]
M_i = -\dfrac{1}{\rho_{n-i}} M_{i-1}, & i = j+2,...,n-1 \\[2ex]
M_j = -\dfrac{6}{h^2} \cdot \dfrac{2 + \dfrac{1}{\rho_{j-1}} + \dfrac{1}{\rho_{n-j-1}}}{4 - \dfrac{1}{\rho_{j-1}} + \dfrac{1}{\rho_{n-j-1}}} & j \neq 0, 1, n-1, n \\[3ex]
M_{j-1} = \dfrac{1}{\rho_{j-1}}\left(\dfrac{6}{h^2} - M_j\right) & j \neq 0, 1, n-1, n \\[2ex]
M_{j+1} = \dfrac{1}{\rho_{n-j-1}}\left(\dfrac{6}{h^2} - M_j\right) & j \neq 0, 1, n-1, n
\end{cases}
$$

where ρ_i are defined recursively;

$$\rho_1 = 4, \quad \rho_i = 4 - \frac{1}{\rho_{i-1}}, \quad i = 2,\ldots$$

Check that following inequalities hold:

$$4 = \rho_1 > \rho_2 > \cdots > \rho_i > \rho_{i+1} > 2 + \sqrt{3} > 3.7 \quad 0.25 < \frac{1}{\rho_i} < 0.3$$

Solution: as S_j s are natural spline functions, so $M_0 = M_n = 0$ and $\lambda_0 = \mu_n = 0$ and $d_0 = d_n = 0$. Assume that j is constant, then we put

$$Y = \{0,\ldots,0,1,0,\ldots0\}$$

where $y_j = 1$. Because the points are equidistant, then $h_{i+1} = h_i = h$, therefore, in the following system:

$$\mu_i M_{i-1} + 2M_i + \lambda_i M_{i+1} = d_i, \quad i = 1,\ldots,n-1$$

the equations:

$$\lambda_i = \frac{h_{i+1}}{h_i + h_{i+1}}$$

$$\mu_i = 1 - \lambda_i$$

$$d_i = \frac{6}{h_i + h_{i+1}} \left[\frac{y_{i+1} - y_i}{h_{i+1}} - \frac{y_i - y_{i-1}}{h_i} \right]$$

are summarized as follows:

$$\mu_i = \lambda_i = \frac{1}{2}, \quad i = 1,\ldots,n-1$$

$$d_i = \begin{cases} \dfrac{3}{h^2}, & i = j-1, j+1 \\[2mm] -\dfrac{6}{h^2}, & i = j \\[2mm] 0, & \text{o.w.} \end{cases}$$

By substituting above coefficients, the matrix form of the above system is as follows:

$$\begin{pmatrix} 2 & 0 & 0 & 0 & \cdots & 0 & 0 & 0 & 0 & 0 & 0 \\ \frac{1}{2} & 2 & \frac{1}{2} & 0 & 0 & \cdots & 0 & 0 & 0 & 0 & 0 \\ 0 & \frac{1}{2} & 2 & \frac{1}{2} & 0 & 0 & \cdots & 0 & 0 & 0 \\ 0 & 0 & \ddots & \ddots & \ddots & 0 & 0 & 0 & \cdots & 0 & 0 \\ 0 & 0 & 0 & \frac{1}{2} & 2 & \frac{1}{2} & 0 & 0 & 0 & \cdots & 0 \\ \vdots & 0 & 0 & 0 & \frac{1}{2} & 2 & \frac{1}{2} & 0 & 0 & 0 & \vdots \\ 0 & \vdots & 0 & 0 & 0 & \frac{1}{2} & 2 & \frac{1}{2} & 0 & 0 & 0 \\ 0 & 0 & \vdots & 0 & 0 & 0 & \ddots & \ddots & \ddots & 0 & 0 \\ 0 & 0 & 0 & \vdots & 0 & 0 & 0 & \frac{1}{2} & 2 & \frac{1}{2} & 0 \\ 0 & 0 & 0 & 0 & \vdots & 0 & 0 & 0 & \frac{1}{2} & 2 & \frac{1}{2} \\ 0 & 0 & 0 & 0 & 0 & 0 & \cdots & 0 & 0 & 0 & 2 \end{pmatrix} \begin{bmatrix} 0 \\ M_1 \\ M_2 \\ \vdots \\ M_{j-2} \\ M_{j-1} \\ M_j \\ M_{j+1} \\ \vdots \\ M_{n-3} \\ M_{n-2} \\ M_{n-1} \\ 0 \end{bmatrix} = \begin{bmatrix} 0 \\ 0 \\ \vdots \\ 0 \\ \dfrac{3}{h^2} \\ -\dfrac{6}{h^2} \\ \dfrac{3}{h^2} \\ 0 \\ \vdots \\ 0 \\ 0 \end{bmatrix}$$

Then, we will have:

$$(1) = \begin{cases} 2M_1 + \dfrac{1}{2}M_2 & 0 \\ \dfrac{1}{2}M_1 + 2M_2 + \dfrac{1}{2}M_3 & 0 \\ \vdots & \vdots \\ \dfrac{1}{2}M_{j-3} + 2M_{j-2} + \dfrac{1}{2}M_{j-1} & 0 \end{cases}$$

$$(2) = \begin{cases} \dfrac{1}{2}M_{j-2} + 2M_{j-1} + \dfrac{1}{2}M_j & = \dfrac{3}{h^2} \\ \dfrac{1}{2}M_{j-1} + 2M_j + \dfrac{1}{2}M_{j+1} & = -\dfrac{6}{h^2} \\ \dfrac{1}{2}M_j + 2M_{j+1} + \dfrac{1}{2}M_{j+2} & = \dfrac{3}{h^2} \end{cases}$$

$$(3) = \begin{cases} \dfrac{1}{2}M_{j+1} + 2M_{j+2} + \dfrac{1}{2}M_{j+3} & = 0 \\ \qquad\qquad \vdots & \vdots \\ \dfrac{1}{2}M_{n-3} + 2M_{n-2} + \dfrac{1}{2}M_{n-1} & = 0 \\ \dfrac{1}{2}M_{n-2} + 2M_{n-1} & = 0 \end{cases}$$

From the first equation of the system, we have:

$$M_1 = -\frac{1}{4}M_2 \Rightarrow M_1 = -\frac{1}{\rho_1}M_2$$

From the second equation, we obtain:

$$M_3 = -\left(4 - \frac{1}{\rho_1}\right)M_2 \Rightarrow M_2 = -\frac{1}{\rho_2}M_3$$

And by continuing this trend, it can be concluded that:

$$M_{j-2} = -\frac{1}{\rho_{j-2}}M_{j-1}$$

Therefore, the following equation generally holds:

$$M_i = -\frac{1}{\rho_i}M_{i+1}, \quad i = 1, \ldots, j-2$$

From the system of the equations (3), we obtain:

$$M_{n-1} = -\frac{1}{4}M_{n-2} = -\frac{1}{\rho_1}M_{n-2} = -\frac{1}{\rho_{n-(n-1)}}M_{n-2}$$

and

$$\frac{1}{2}M_{n-3} + 2M_{n-2} + \frac{1}{2}M_{n-1} = 0$$

Therefore,

$$M_{n-2} = -\left(\frac{1}{4 - \dfrac{1}{\rho_1}}\right)M_{n-3} = -\frac{1}{\rho_2}M_{n-3} = -\frac{1}{\rho_{n-(n-2)}}M_{n-3}$$

And by continuing this trend, we will have:

$$M_{j+2} = -\frac{1}{\rho_n - (j+2)} M_{j+1}$$

And it can generally be said that:

$$M_i = -\frac{1}{\rho_{n-i}} M_{i-1}, \quad i = j+2,\ldots,n-1$$

According to the first equation of the system (2), we have:

$$\frac{1}{2} M_{j-2} + 2M_{j-1} + \frac{1}{2} M_j = \frac{3}{h^2}$$

Hence,

$$\frac{1}{2}\left(-\frac{1}{\rho_{j-2}}\right) M_{j-1} + 2M_{j-1} = \frac{3}{h^2} - \frac{1}{2} M_j$$

So,

$$\left(4 - \frac{1}{\rho_{j-2}}\right) M_{j-1} - \frac{6}{h^2} - M_j$$

Consequently,

$$\left(4 - \frac{1}{\rho_{j-2}}\right) M_{j-1} = \frac{6}{h^2} - M_j$$

Using the third equation of the system (2), we have:

$$\frac{1}{2} M_j + 2M_{j+1} + \frac{1}{2} M_{j+2} = \frac{3}{h^2}$$

Therefore,

$$\frac{1}{2} M_j + 2M_{j+1} + \frac{1}{2}\left(-\frac{1}{\rho_{n-j-2}}\right) M_{j+1} = \frac{3}{h^2}$$

Then,

$$\left(4 - \frac{1}{\rho_{n-j-2}}\right) M_{j+1} = \frac{6}{h^2} - M_j$$

And hence,

$$M_{j+1} = \frac{1}{\rho_{n-j-1}}\left(\frac{6}{h^2} - M_j\right)$$

And also, using the second equation of the system (2), we have:

$$\frac{1}{2}M_{j-1} + 2M_j + \frac{1}{2}M_{j+1} = -\frac{6}{h^2}$$

Consequently,

$$\frac{1}{2}\left(\frac{1}{\rho_{j-1}}\left(\frac{6}{h^2} - M_j\right)\right) + 2M_j + \frac{1}{2}\left(\frac{1}{\rho_{n-j-1}}\left(\frac{6}{h^2} - M_j\right)\right) = -\frac{6}{h^2}$$

In this case,

$$M_j = -\frac{3}{h^2} - \frac{1}{4}\left(\frac{6}{h^2} - M_j\right)\left(\frac{1}{\rho_{j-1}} + \frac{1}{\rho_{n-j-1}}\right)$$

$$= -\frac{3}{h^2} - \frac{3}{2h^2}\left(\frac{1}{\rho_{j-1}} + \frac{1}{\rho_{n-j-1}}\right) + \frac{1}{4}M_j\left(\frac{1}{\rho_{j-1}} + \frac{1}{\rho_{n-j-1}}\right)$$

$$= -\frac{3}{h^2}\left[1 + \frac{1}{2}\left(\frac{1}{\rho_{j-1}} + \frac{1}{\rho_{n-j-1}}\right)\right] + \frac{1}{4}M_j\left(\frac{1}{\rho_{j-1}} + \frac{1}{\rho_{n-j-1}}\right)$$

And therefore,

$$M_j\left[1 - \frac{1}{4}\left(\frac{1}{\rho_{j-1}} + \frac{1}{\rho_{n-j-1}}\right)\right] = -\frac{3}{h^2}\left[1 + \frac{1}{2}\left(\frac{1}{\rho_{j-1}} + \frac{1}{\rho_{n-j-1}}\right)\right]$$

Subsequently,

$$M_j = -\frac{3}{h^2}\left(\frac{2 + \dfrac{1}{\rho_{j-1}} + \dfrac{1}{\rho_{n-j-1}}}{2}\bigg/\frac{4 - \dfrac{1}{\rho_{j-1}} + \dfrac{1}{\rho_{n-j-1}}}{2}\right) = -\frac{6}{h^2} \cdot \frac{2 + \dfrac{1}{\rho_{j-1}} + \dfrac{1}{\rho_{n-j-1}}}{4 - \dfrac{1}{\rho_{j-1}} + \dfrac{1}{\rho_{n-j-1}}}$$

We prove the second part by contradiction:

$$\rho_i = 4 - \frac{1}{\rho_{i-1}}, \quad i = 2,3,\ldots$$

If $i = 2$, then

$$\rho_2 = 4 - \frac{1}{\rho_1} = 4 - \frac{1}{4} < 4 = \rho_1$$

Suppose that $\rho_i < \rho_{i-1}$, we prove $\rho_{i+1} < \rho_i$

$$\rho_{i+1} = 4 - \frac{1}{\rho_i} < 4 - \frac{1}{\rho_{i-1}} = \rho_i$$

So, always we have:

$$4 = \rho_1 > \rho_2 > \cdots > \rho_i > \rho_{i+1}$$

If we put $S := \rho_{i+1}$, then we have

$$S = 4 - \cfrac{1}{4 - \cfrac{1}{4 - \cfrac{1}{4 - \cfrac{\ddots}{4 - \cfrac{1}{4}}}}} = 4 - \frac{1}{S}$$

So,

$$S^2 - 4S + 1 = 0 \Rightarrow S = 2 \pm \sqrt{3}$$

$$\Rightarrow 0.25 < \frac{1}{\rho_i} < 0.3$$

$$\Rightarrow 4 = \rho_1 > \cdots > \rho_i > \rho_{i+1} > 2 + \sqrt{3} = 3.7$$

4.5.43 Problem

Suppose that S is a linear space consisting of all spline functions S_Δ with the nodes $\Delta = \{x_0,\ldots,x_n\}$ and

$$S_\Delta''(x_0) = S_\Delta''(x_n) = 0$$

and the functions $S_0,\ldots,S_n$ are the spline functions defined in the problem (4.5.42). Show that for $Y = \{y_0,\ldots,y_n\}$, we have:

$$S_\Delta(Y,x) \equiv \sum_{j=0}^{n} y_j S_j(x)$$

and determine the dimension of space S.

Solution: we consider points $x_0,\ldots,x_n$ with intercepts $y_0,\ldots,y_n$ and assume that:

$$S_\Delta(x) = S_i(x), \quad x \in [x_i, x_{i+1}], \quad i = 0,\ldots,n-1$$

We build a set as follows:

Using Lagrange polynomials, we write an interpolation for data of Table 4.3 and name it $G_0(x)$ so that $G_0 \in \Pi_n$, and

$$G_0(x) = \sum_{i=0}^{n} L_i(x) g_i = L_0(x)$$

and if we continue this procedure, for j-th table, we have (Table 4.4):

$$G_j(x) = \sum_{i=0}^{n} L_i(x) g_i = L_j(x)$$

and finally, for n-th tale, we have:

$$G_n(x) = \sum_{i=0}^{n} L_i(x) g_i = L_n(x)$$

So, it can be concluded that:

TABLE 4.3

Data for 4.5.43

x_i	x_0	x_1	...	x_n
g_i	1	0	...	0

TABLE 4.4

Data for Problem 4.5.43

x_i	x_0	...	x_{j-1}	x_j	x_{j+1}	...	x_n
g_i	0	...	0	1	0	...	0

$$S(x) = \sum_{i=0}^{n} G_i(x) y_i, \quad x \in [x_i, x_{i+1}]$$

where

$$S(x_i) = \sum_{i=0}^{n} G_i(x_i) y_i = y_i$$

So, the interpolation condition holds. Now, if we put $S_j(x) = G_j(x)$, then

$$S_\Delta(x) = \sum_{j=0}^{n} S_j(x) y_j$$

Since $S_j(x)$ are polynomials of degree n and S_Δ is the linear combination of S_j s, then the basis of space, that is, set $\{1, x, x^2, \ldots, x^n\}$ consists of polynomials of at most degree n. Therefore, the dimension of space is equal to $n+1$.

4.5.44 Problem

Bessel function of degree zero as;

$$J_0(x) = \frac{1}{\pi} \int_0^\pi \cos(x \sin t) dt$$

is tabulated in equal arguments as $x_i = x_0 + ih$, where $i = 0, 1, \ldots, n$.

a. If the linear interpolator is used, how small should the increment h be chosen so that the interpolation error is less that 10^{-6}?

b. When n tend to infinity, how does the maximal interpolation error behave? In the above relation, $p_n(x)$ interpolates $J_0(n)$ in the points $x = x_i^n := \dfrac{i}{n}$.

c. If $S_{\Delta_n}(x)$ are spline interpolating functions with nodes $\Delta_n = \{x_i^{(n)}\}$ and $S_{\Delta_n}(x) = J_0'(x)$ for $x = 0, 1$, compare above result with the error behavior $\max\limits_{0 \le x \le 1} |S_{\Delta_n}(x) - J_0(x)|$ as $n \to \infty$.

Solution: If

$$w(x) = (x - x_0) \ldots (x - x_n)$$

Then,

$$|R(x)| = |p_n(x) - J_0(x)| = \left| \frac{w(x)}{(n+1)!} J_0^{(n+1)}(x) \right|$$

Therefore,

$$\max_{0 \le x \le 1} |R(x)| = \max_{0 \le x \le 1} |p_n(x) - J_0(x)|$$

$$\le \frac{|w(x)|}{(n+1)!} \max_{0 \le x \le 1} |J_0^{(n+1)}(x)|$$

In order to calculate the upper bound of error, we should obtain a bound for $\max_{0 \le x \le 1} |J_0^{(n+1)}(x)|$. To do this, we first obtain $J_0^{(n+1)}(x)$:

$$J_0'(x) = \frac{d}{dx} J_0(x) = \frac{1}{\pi} \int_0^\pi \frac{d}{dx} \cos(x \sin t) dt$$

$$= \frac{1}{\pi} \int_0^\pi -\sin t \cdot \sin(x \sin t) dt$$

$$= \frac{1}{\pi} \int_0^\pi -\sin t \cdot \left[-\cos\left(\frac{\pi}{2} + x \sin t \right) \right] dt$$

$$= \frac{1}{\pi} \int_0^\pi \sin t \cdot \cos\left(\frac{\pi}{2} + x \sin t \right) dt$$

Hence, according to the above equations, it can be concluded that:

$$J_0^{(k)}(x) = \frac{1}{\pi} \int_0^\pi \sin^k t \cdot \cos\left(\frac{k\pi}{2} + x \sin t \right) dt$$

So,

$$\left| J_0^{(n+1)}(x) \right| \le \frac{1}{\pi} \int_0^\pi |\sin^{n+1} t| \cdot \left| \cos\left(\frac{(n+1)\pi}{2} + x \sin t \right) \right| dt$$

$$\le \frac{1}{\pi} \int_0^\pi dt$$

$$= 1$$

So, the upper bound of error is obtained as follows:

$$\max_{0 \le x \le 1} |p_n(x) - J_0(x)| \le \frac{|w(x)|}{(n+1)!} \max_{0 \le x \le 1} |J_0^{(n+1)}(x)| \le \frac{|w(x)|}{(n+1)!}$$

a. If $x = x_0 + \theta h$, then $x - x_i = (\theta - i)h$ and therefore, $w(x)$ is obtained as follows:

$$w(x) = (x - x_0)...(x - x_n)$$

$$= \theta h \cdot (\theta - 1)h...(\theta - n)h$$

$$= \theta \cdot (\theta - 1)...(\theta - n) \cdot h^{n+1}$$

Hence,

$$\max_{0 \le x \le 1} |p_n(x) - J_0(x)| \le \frac{|\theta \cdot (\theta - 1)...(\theta - n) \cdot h^{n+1}|}{(n+1)!} \le \frac{\theta^{n+1} \cdot h^{n+1}}{(n+1)!}$$

Now, if $\dfrac{\theta^{n+1} \cdot h^{n+1}}{(n+1)!} \le 10^{-6}$, then $h \le \dfrac{\sqrt[n+1]{10^{-6}(n+1)!}}{\theta}$.

b. Given $h = \dfrac{1}{n}$, we have

$$\max_{0 \le x \le 1} |p_n(x) - J_0(x)| \le \frac{\theta^{n+1}}{n^{n+1}(n+1)!}$$

When n tends to infinity, the maximal error behaves as follows:

$$\lim_{n \to \infty} \max_{0 \le x \le 1} |p_n(x) - J_0(x)| \le \lim_{n \to \infty} \frac{\theta^{n+1}}{n^{n+1}(n+1)!} = 0$$

c. According to the spline convergence theorem in, we have:

$$|S_\Delta(x) - f(x)| \le C_k L \|\Delta\|^4$$

where

$$C_k < 2, \quad |f^{(4)}(\xi)| \le L, \quad \Delta = \max_{j=0,1,...,n-1} |x_{j+1} - x_j|$$

Therefore,

$$|S_\Delta(x) - J_0(x)| \le C_k \cdot 1 \cdot h^4$$

Now, given that $h = \dfrac{1}{n}$, we have:

$$|S_\Delta(x) - J_0(x)| \le C_k \cdot \frac{1}{n^4} < \frac{2}{n^4}$$

Therefore,

$$\lim_{n\to\infty}\left|S_\Delta(x)-J_0(x)\right|<\lim_{n\to\infty}\frac{2}{n^4}=0$$

4.5.45 Problem

Let $\Delta=\{x_0<\cdots<x_n\}$. Show that the spline function S_Δ with the boundary conditions:

$$S_\Delta^{(k)}(x_0)=S_\Delta^{(k)}(x_n)=0,\quad k=0,1,2$$

for $n<4$ is equal to zero.

Solution: according to the above boundary conditions, it is enough to prove that for $n=1,2,3$ and $j=0,1,2,\ldots,n-1$, the coefficients $\alpha_j,\ \beta_j\ ,\gamma_j,\ \delta_j$ of the following spline equation are always equal to zero:

$$S_\Delta(x)=\alpha_j+\beta_j(x-x_j)+\gamma_j(x-x_j)^2+\delta_j(x-x_j)^3,\quad x\in[x_j,x_{j+1}]$$

Suppose that $n=1$. Using the boundary conditions, we form a system of equations with six equations and six unknowns.

$$\begin{cases} S_\Delta(x_0)=S_\Delta(x_1)=0 \\ S_\Delta'(x_0)=S_\Delta'(x_1)=0 \\ S_\Delta''(x_0)=S_\Delta''(x_1)=0 \end{cases}$$

The matrix form of above system is as follows:

$$\begin{bmatrix} 1 & (x_0-x_j) & (x_0-x_j)^2 & (x_0-x_j)^3 \\ 1 & (x_1-x_j) & (x_1-x_j)^2 & (x_1-x_j)^3 \\ 0 & 1 & 2(x_0-x_j) & 3(x_0-x_j)^2 \\ 0 & 1 & 2(x_1-x_j) & 3(x_1-x_j)^2 \\ 0 & 0 & 2 & 6(x_0-x_j) \\ 0 & 0 & 2 & 6(x_1-x_j) \end{bmatrix} \begin{bmatrix} \alpha_j \\ \beta_j \\ \gamma_j \\ \delta_j \end{bmatrix} = \begin{bmatrix} 0 \\ 0 \\ 0 \\ 0 \end{bmatrix}$$

The matrix of above system's coefficients is 6×4, so it can be rewritten in the following form:

$$\left[\begin{matrix} B \\ C \end{matrix} \right] X = 0$$

where B is a 4×4 non-singular matrix and C is a 2×2 matrix. Therefore,

$$\begin{cases} BX = 0 \\ CX = 0 \end{cases}$$

4.5.46 Problem

Interpolation in the product space: suppose that every linear interpolating problem expressed in terms of functions $\varphi_0, \ldots, \varphi_n$ for distinct nodal points $x_0, \ldots, x_n$ has a unique answer as follows:

$$\phi(x) = \sum_{i=0}^{n} \alpha_i \varphi_i(x)$$

where

$$\phi(x_k) = f_k, \quad k = 0, \ldots, n$$

Prove that if $\{\psi_0, \ldots, \psi_m\}$ is a set of functions that has a unique answer for each linear interpolation problem, then for every choice of distinct points with lengths $x_0, \ldots, x_n$ and $y_0, \ldots, y_n$ and width f_{ik}, $i = 0, \ldots, n, k = 0, \ldots, m$, there is a unique function such as

$$\phi(x, y) = \sum_{v=0}^{n} \sum_{\mu=0}^{m} \alpha_{v\mu} \varphi_v(x) \psi_\mu(y)$$

where for $i = 0, \ldots, n$ and $k = 0, \ldots, m$:

$$\phi(x_i, y_k) = f_{ik}$$

Solution: for an arbitrary and hereafter constant k and points (x_i, f_i) with $i = 0, \ldots, n$, there is a unique function of $\phi_k(x)$ such that:

$$\phi_k(k) = \sum_{i=0}^{n} \alpha_i(y_k) \varphi_i(x)$$

$$\phi_k(x_i) = f_{ik}, \quad i = 0, \ldots, n$$

So, we can write:

$$k = 0: \quad \phi_0(x) = \alpha_0(y_0)\varphi_0(x) + \alpha_1(y_0)\varphi_1(x) + \cdots + \alpha_n(y_0)\varphi_n(x)$$

$$k = 0: \quad \phi_1(x) = \alpha_0(y_1)\varphi_0(x) + \alpha_1(y_1)\varphi_1(x) + \cdots + \alpha_n(y_1)\varphi_n(x)$$

$$\vdots \qquad\qquad\qquad\qquad\qquad \vdots$$

$$k = m: \quad \phi_m(x) = \alpha_0(y_m)\varphi_0(x) + \alpha_1(y_m)\varphi_1(x) + \cdots + \alpha_n(y_m)\varphi_n(x)$$

Suppose that $\alpha_i(y_k)$ be a function of y_k so that for $m+1$ points of y_k has the following form:

y_k		y_0	y_1	$\cdots$	y_m
$\alpha_i(y_k)$		$\alpha_i(y_0)$	$\alpha_i(y_1)$	$\cdots$	$\alpha_i(y_m)$

So, using interpolation, we obtain $\alpha_i(y)$ as follows:

$$\alpha_i(y) = \sum_{\mu=0}^{m} \beta_{i_\mu}\psi_\mu(y) + R_i(y), \quad i = 0,\ldots,n$$

where

$$R_i(y) = \frac{(y - y_0)\cdots(y - y_m)}{(m+1)!}\alpha_i^{(m+1)}(y) \quad i = 0,\ldots,n$$

$$R_i(y_k) = 0 \qquad\qquad\qquad k = 0,\ldots,m$$

Now we have:

$$\phi_y(x) = \sum_{i=0}^{n}\alpha_i(y)\varphi_i(x)$$

$$= \sum_{i=0}^{n}\left(\sum_{\mu=0}^{m}\beta_{i_\mu}\psi_\mu(y) + R_i(y)\right)\varphi_i(x)$$

$$= \sum_{i=0}^{n}\sum_{\mu=0}^{m}\beta_{i_\mu}\psi_\mu(y)\varphi_i(x) + \sum_{i=0}^{n}R_i(y)\varphi_i(x)$$

$$= \phi(x,y) + G(x,y)$$

where $G(x,y) = \sum_{i=0}^{n}R_i(y)\varphi_i(x)$, then $\phi(x,y)$ is uniquely obtained and for $i = 0,\ldots,n$ and $k = 0,\ldots,m$, we have:

$$\phi\left(x_i, y_k\right) = \sum_{v=0}^{n} \sum_{\mu=0}^{m} \beta_{v_\mu} \varphi_v\left(x_i\right) \psi_\mu\left(y_k\right) + 0 = \sum_{v=0}^{n} \alpha_v\left(y_k\right) \varphi_v\left(x_i\right) = f_{ik}$$

4.6 Reciprocal Interpolation

So far, we have estimated the value of $f(x)$ for a given x. If we mean to estimate a value of x so that $f(x)$ has a known value, this is called reciprocal interpolation. Reciprocal interpolation also has applications. For example, the roots of $f(x) = 0$ can be estimated using it, so that we get an x for which $f(x)$ is equal to zero. Also, if we have a population on a certain time interval and we want to determine the approximate year at which, the population has a certain number, we use inverse interpolation.

4.6.1 Transforming Reciprocal Interpolation to Direct Interpolation

If $y = f(x)$ and f have inverse functions, we have

$$x = f^{-1}(y)$$

So, instead of Table 4.5,
  Table 4.6 can be considered
  Given that the f_is are usually not equidistant and we do not have a choice to select them, then Newton's forward and backward formulas cannot be used. Hence, the only effective method that can always be used is the Newton's divided difference table, in which the roles of x_i and f_i are exchanged.

TABLE 4.5

Data for 4.6.1

x_i	x_0	x_1	...	x_n
f_i	f_0	f_1	...	f_n

TABLE 4.6

Data for 4.6.1

f_i	f_0	f_1	...	f_n
x_i	x_0	x_1	...	x_n

4.6.2 Example

Consider the following Table 4.7 of function $f(x) = \sin x$.

We want to determine x such that $f(x) = 0.2$. For this purpose, we form the table of the Newton's divided differences as follows (Table 4.8):

Using Newton's interpolating polynomial, we can write:

$$x \approx x_0 + (f - f_0) f[f_0, f_1] + \cdots + (f - f_0) \ldots (f - f_{n-1}) f[f_0, \ldots, f_n]$$

where

$$f[f_0, f_1] = \frac{x_1 - x_0}{f_1 - f_0}$$

$$f[f_0, \ldots, f_n] = \frac{f[f_1, \ldots, f_n] - f[f_0, \ldots, f_{n-1}]}{f_n - f_0}$$

Thus,

$$x \approx 10 + (0.2 - 0.1735)59.38 + (0.2 - 0.1736)(0.2 - 0.3420)5.18$$

$$+ \cdots + (0.2 - 0.1734) \ldots (0.2 - 0.6428)34.86$$

$$= 10 + 1.5676 - 0.0194 - 0.0102 + 0.0030 - 0.0035$$

$$= 11.5375$$

TABLE 4.7

Interpolation Data for $f(x) = \sin(x)$

x_i	0°	10°	20°	30°	40°	50°
f_i	0	0.1736	0.3420	0.5	0.6428	0.7660

TABLE 4.8

Table of Newton's Divided Differences for Example 4.6.2

f_i	x_i	$f[f_0, f_1]$	$f[f_0, f_1, f_2]$	$f[f_0, \ldots, f_3]$	$f[f_0, \ldots, f_4]$	$f[f_0, \ldots, f_5]$
0.1736	10	59.38				
0.3420	20	58.48	5.18			
0	0	60.00	9.62	19.88	13.38	34.86
0.5	30	70.03	15.60	34.31		
0.6428	40	81.17	41.88			
0.766	50					

So, the answer is about 11.5375 degrees. To examine the correctness of the answers, it is enough to estimate $f(11.5375)$, if it has a value close to 0.2, the answer is correct.

4.7 Exercise

1. Find the Hermit interpolation for the function $f(x) = \sqrt{x}$ on the points $x_0 = 1, x_1 = \dfrac{3}{2}, x_2 = 2$.

2. If $p_n(x)$ be polynomial interpolation on the points $x_0, x_0 + \in, \dots, x_n$, then check

$$\lim_{\varepsilon \to 0} p_n(x)$$

3. Find the B-spline interpolation of $\{(-3,3), (-1,1), (0,0), (1,1), (3,3)\}$

4. If $f'(x) = \dfrac{df(x)}{dx}$, show that

$$\frac{d}{dx} f[x_0, x] \neq f'[x_0, x]$$

unless $f(x)$ be linear.

5. If $f(x) = \dfrac{ax+b}{cx+d}$ and $x \neq y$, find

$$f[x,y], \quad f[x,x,y], \quad f[x,x,y,y]$$

6. Suppose that $f(x) = x^5$, then if $a \neq b \neq c$, find

$$f[a,b,c], \quad f[a,a,b], \quad f[a,a,a]$$

7. Show that:

$$f[x_v, \dots, x_{v+l+1}] = \frac{1}{l!(x_{v+l+1} - x_v)} \int_R B_{lv} f^{(l+1)}(x) dx$$

8. Suppose $f(x) = \dfrac{1}{a-x}$. If $x, x_n, \dots, x_0 \neq a$, show that:

A.

$$f[x_0,\ldots,x_n] = \frac{1}{(a-x_0)\ldots(a-a_n)}$$

B.

$$f[x_0,\ldots,x_n,x] = \frac{1}{(a-x_0)\ldots(a-x_n)(a-x)}$$

5

Interval Interpolation

5.1 Interval Interpolation

Sometimes the points used for interpolation may not be exactly available and we may have some parts of a data. For this type of data, the word "approximately" is commonly used. Now, if we want to make the word approximately and about meaningful, one of the ways is to use the interval, for example, we show about 2 with $[2-\varepsilon, 2+\varepsilon]$, where ε is introduced arbitrarily. So, instead of about 2 or approximately 2, you can use a range that contains 2. Classification of interval data is possible in two ways.

1. Data whose length and width are interval.
2. Data that only whose width is an interval.

Checking case (1) is more difficult than checking case (2). Because in case (1), the resulting interpolation criterion will have a complex and problematic form, so we will ignore it and only examine case (2). In this case, it is assumed that $n+1$ points are as follows:

$$\left(x_i, \left[\underline{f_i}, \overline{f_i}\right]\right), \quad i=0,\ldots,n$$

If we want to consider the interpolation problem

$$\phi(x; a_0,\ldots,a_n) = \sum_{j=0}^{n} a_j \phi_j(x) \tag{5.1}$$

Such that the interpolation condition is satisfied, we have the following interval data

$$\phi(x; a_0,\ldots,a_n) = \left[\underline{f_i}, \overline{f_i}\right], \quad i=0,\ldots,n$$

DOI: 10.1201/9781003218173-5

According to the right side of the equation, it is obvious that the left side must also be an interval. Therefore, it can be written

$$\left[\underline{\phi}_i, \overline{\phi}_i\right] = \left[\underline{f}_i, \overline{f}_i\right], \quad i = 0, \ldots, n$$

where

$$\underline{\phi} = \min\left\{u \mid \underline{\phi} \leq u \leq \overline{\phi}\right\}$$
$$\overline{\phi} = \max\left\{u \mid \underline{\phi} \leq u \leq \overline{\phi}\right\}$$

So, according to the equality between two intervals, we have:

$$L : \underline{\phi}(x_i; a_0, \ldots, a_n) = \underline{f}_i$$
$$U : \overline{\phi}(x_i; a_0, \ldots, a_n) = \overline{f}_i, \quad i = 0, \ldots, n$$

So, for every i, we have two interpolation problems of L and U. So totally, we will have two system of order $n + 1$. According to the equation (5.1), $\underline{\phi}$ can be introduced as follows:

$$\underline{\phi}(x; a_0, \ldots, a_n) = \sum_{j=0}^{n} \underline{a_j \phi_j(x)} = \sum_{j=0}^{n} \underline{a_j \phi_j(x)}$$

By multiplying a scalar to the interval, it can be claimed that:

$$\underline{a_j \phi_j(x)} = \begin{cases} \underline{a_j} \phi_j(x), & \phi_j(x) \geq 0 \\ \overline{a_j} \phi_j(x), & \phi_j(x) < 0 \end{cases}$$

Therefore, it can be written

$$\underline{\phi}(x; a_0, \ldots, a_n) = \sum_{\phi_j(x) \geq 0} \underline{a_j} \phi_j(x) + \sum_{\phi_j(x) < 0} \overline{a_j} \phi_j(x) \tag{5.2}$$

So interpolation problem (L) is obtained as follows:

$$\underline{\phi}(x_i; a_0, \ldots, a_n) = \sum_{\phi_j(x_i) \geq 0} \underline{a_j} \phi_j(x_i) + \sum_{\phi_j(x_i) < 0} \overline{a_j} \phi_j(x_i) = \underline{f}_i, \quad i = 0, \ldots, n$$

It can also be said that

$$\overline{\phi}(x;a_0,\ldots,a_n)=\sum_{j=0}^{n}\overline{a_j\phi_j(x)}=\sum_{j=0}^{n}\overline{a_j\phi_j(x)}$$

and as before

$$\overline{a_j\phi_j(x)}=\begin{cases}\overline{a_j}\phi_j(x), & \phi_j(x)\geq 0\\ \underline{a_j}\phi_j(x), & \phi_j(x)<0\end{cases}$$

So, we will have

$$\overline{\phi}(x;a_0,\ldots,a_n)=\sum_{\phi_j(x)\geq 0}\overline{a_j}\phi_j(x)+\sum_{\phi_j(x)<0}\underline{a_j}\phi_j(x) \tag{5.3}$$

and the interpolation problem (U) is as follows:

$$\overline{\phi}(x_i;a_0,\ldots,a_n)=\sum_{\phi_j(x_i)\geq 0}\overline{a_j}\phi_j(x_i)+\sum_{\phi_j(x_i)<0}\underline{a_j}\,\phi_j(x_i)=\overline{f_i},\quad i=0,\ldots,n$$

as a result

$$L:\sum_{\phi_j(x_i)\geq 0}\underline{a_j}\phi_j(x_i)+\sum_{\phi_j(x_i)<0}\overline{a_j}\phi_j(x_i)=\underline{f_i}$$

$$U:\sum_{\phi_j(x_i)\geq 0}\overline{a_j}\phi_j(x_i)+\sum_{\phi_j(x_i)<0}\underline{a_j}\phi_j(x_i)=\overline{f_i} \qquad i=0,\ldots,n \tag{5.4}$$

According to the equations (5.2) and (5.4), it can be said that $\overline{\phi}$ and ϕ form an interval. So, in this case, we always have two types of interpolation problems as (L) and (U), each of which must uniquely exist, and (L) and (U) interpolate points $\left(x_i, \underline{f_i}\right)$ and $\left(x_i, \overline{f_i}\right)$, respectively. So, in this case, instead of an interpolating function, we have a band of an interpolating function (Figure 5.1).

Therefore, it is necessary for each group of the points $\left(x_i, \underline{f_i}\right)$ and $\left(x_i, \overline{f_i}\right)$, $i=0,\ldots,n$ to form a function, and an interpolation problem corresponded to each of them is introduced.

Now if $\phi_j(x)=x^j$, the interpolation interval is as follows:

$$L:\ \underline{p}(x)=\sum_{x^j\geq 0}\underline{a_j}x^j+\sum_{x^j<0}\overline{a_j}x^j$$

$$U:\overline{p}(x)=\sum_{x^j\geq 0}\overline{a_j}x^j(x_i)+\sum_{x^j<0}\underline{a_j}x^j(x_i) \tag{5.5}$$

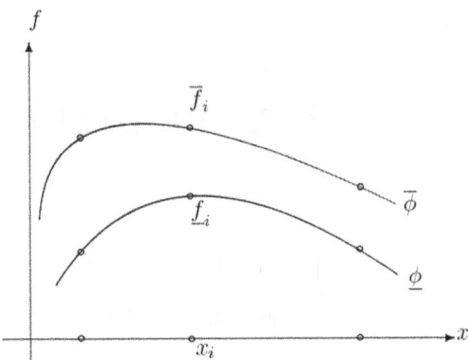

FIGURE 5.1   Interval-valued function.

It is clear that by considering the j th power of x, as even and odd, we can write:

$$L: \underline{p}(x) = \sum_{j=2k} \underline{a}_j x^j + \sum_{j=2k+1} \overline{a}_j x^j \quad x < 0$$

$$\underline{p}(x) = \sum_{j=0}^{n} \underline{a}_j x^j \quad x \geq 0$$

$$k = 0, \ldots, \left[\frac{n}{2} \right]$$

$$U: \overline{p}(x) = \sum_{j=2k} \overline{a}_j x^j + \sum_{j=2k+1} \underline{a}_j x^j, \quad x < 0$$

$$\overline{p}(x) = \sum_{j=0}^{n} \overline{a}_j x^j, \quad x \geq 0$$

$$k = 0, \ldots, \left[\frac{n}{2} \right]$$

The following cases can be considered for system (5.5):

1. If for every x and every j, x^j is non-negative, then,

$$L: \underline{p}(x) = \sum_{j=0}^{n} \underline{a}_j x^j$$

$$U: \overline{p}(x) = \sum_{j=0}^{n} \overline{a}_j x^j$$

(5.6)

2. If for every x and every j, x^j is negative (although this never happens due to a constant multiplication), then we have

$$L : \underline{p}(x) = \sum_{j=0}^{n} \overline{a}_j x^j$$

$$U : \overline{p}(x) = \sum_{j=0}^{n} \underline{a}_j x^j \tag{5.7}$$

3. If for some x and j, x^j is non-negative and for some x and j, x^j are negative, then the system (5.5) is established in the same way.

In case 1, (L) and (U) systems always can be separately solved and also these two systems can be written as a combination of themselves with double dimensions and examines as follows:

$$\begin{bmatrix} B & 0 \\ 0 & B \end{bmatrix} \begin{bmatrix} \underline{a} \\ \overline{a} \end{bmatrix} = \begin{bmatrix} \underline{f} \\ \overline{f} \end{bmatrix}$$

where

$$B = \left[x_i^j \right]_{i,j=0}^{n}$$

$$\underline{a} = \left[\underline{a}_0, \dots, \underline{a}_n \right]^t, \quad \overline{a} = \left[\overline{a}_0, \dots, \overline{a}_n \right]$$

$$\underline{f} = \left[\underline{f}_0, \dots, \underline{f}_n \right]^t, \quad \overline{f} = \left[\overline{f}_0, \dots, \overline{f}_n \right]^t$$

In case (2), (L) and (U) systems always can still be separately solved. But with double dimensions, the combination is as follows:

$$\begin{bmatrix} 0 & C \\ C & 0 \end{bmatrix} \begin{bmatrix} \underline{a} \\ \overline{a} \end{bmatrix} = \begin{bmatrix} \underline{f} \\ \overline{f} \end{bmatrix}$$

where $\underline{a}$, $\overline{a}$, $\underline{f}$, and $\overline{f}$ are the same as the above vectors, and matrix C is defined similar to matrix B. In case (3), if we want to solve systems (L) and (U) separately, then we either have no answer or have infinite answer, because, the number of unknowns exceeds the number of constraints. So, we have to solve the combined system of double order, which is obtained as follows:

$$\begin{bmatrix} B & C \\ C & B \end{bmatrix} \begin{bmatrix} \underline{a} \\ \overline{a} \end{bmatrix} = \begin{bmatrix} \underline{f} \\ \overline{f} \end{bmatrix} \tag{5.8}$$

where $B \geq 0$ and $C \leq 0$.

In all three cases of (1), (2), and (3), it can be claimed that the matrix form is as (5.8). So, assuming

$$S = \begin{bmatrix} B & C \\ C & B \end{bmatrix}, \quad X = \begin{bmatrix} a \\ a \end{bmatrix}, \quad Y = \begin{bmatrix} f \\ f \end{bmatrix}$$

5.1.1 Theorem

$$\det(S) = \det(B - C) \cdot \det(B + C)$$

Proof: If we subtract the last $n+1$ rows from the first $n+1$ rows, respectively, we have

$$\det(S) = \det \begin{bmatrix} B - C & C - B \\ C & B \end{bmatrix}$$

And now if we add the first $n+1$ columns to the second $n+1$ columns, respectively, we have

$$\det(S) = \det \begin{bmatrix} B - C & 0 \\ C & B + C \end{bmatrix}$$

where according to the Laplace expansion of determinant with respect to the first $n+1$ columns, the sentence is true.

5.1.2 Corollary

The necessary and sufficient condition for $\det(S) \neq 0$ is $\det(B - C) \neq 0$, that is, the necessary and sufficient condition for system $SX = Y$ to always have a unique answer is that the matrix $B - C$ is non-singular.

So, unlike ordinary interpolation, it cannot always be said that this type of interpolation is always unique, and the necessary and sufficient condition for it to be unique is that $\det(B - C) \neq 0$. Considering that if S is non-singular, the answer of system $SX = Y$ is obtained as $X = S^{-1}Y$, so it is worth to study the structure of S^{-1}.

5.1.3 Theorem

Matrix S^{-1}, if any, has a structure similar to S that is as follows:

$$S^{-1} = \begin{bmatrix} D & E \\ E & D \end{bmatrix} \tag{5.9}$$

where

$$D = \frac{1}{2}\left((B+C)^{-1} + (B-C)^{-1}\right)$$

$$E = \frac{1}{2}\left((B+C)^{-1} - (B-C)^{-1}\right)$$

(5.10)

Proof: Suppose s_{ij} and t_{ij} are the elements corresponding to the i-th row and j-th column of matrices S and S^{-1}, respectively. Then,

$$t_{ij} = \frac{(-1)^{i+j}|S_{ji}|}{|S|}$$

where S_{ji} is the matrix resulting from the removal of the jth row and the ith column of the matrix S. Now for $1 \le i, j \le n$, we consider elements $t_{i,n+j}$ and $t_{n+i,j}$ of the matrix S^{-1}. If we assume that the matrix $S_{n+j,i}$ is the matrix resulting from the removal of the $(n+j)$th row and the ith column of the matrix S and the matrix $S_{j,n+i}$ is the matrix resulting from the removal of the jth row and the $(n+i)$th column of the matrix S, we can easily show that $S_{n+j,i}$ is obtained by substituting an even number of rows and columns of $S_{j,n+i}$. Thus,

$$t_{i,n+j} = (-1)^{i+n+j}\frac{|S_{n+j,i}|}{|S|}$$

$$= (-1)^{i+n+j}\frac{|S_{j,n+i}|}{|S|} = t_{n+i,j}$$

Similarly, for arbitrary i and j, it is proved that $t_{ij} = t_{n+i,n+j}$. As a result, we can say that S^{-1} has a structure as (5.9).

Now we show that the equation (5.9) hold. We have

$$SS^{-1} = \begin{bmatrix} B & C \\ C & B \end{bmatrix}\begin{bmatrix} D & E \\ E & D \end{bmatrix} = \begin{bmatrix} I & 0 \\ 0 & I \end{bmatrix}$$

Thus,

$$BD + CE = I, \quad CD + BE = 0$$

(5.11)

If we first add and then subtract the equation (5.10), respectively, we will have

$$D + E = (B+C)^{-1} \quad D - E = (B-C)^{-1}$$

and as a result

$$D = \frac{1}{2}\left((B+C)^{-1} + (B-C)^{-1}\right)$$

$$E = \frac{1}{2}\left((B+C)^{-1} - (B-C)^{-1}\right)$$

5.1.4 Point

$$\begin{bmatrix} B_{\geq 0} & C \\ C_{\leq 0} & B \end{bmatrix} \begin{bmatrix} \underline{X} \\ \overline{X} \end{bmatrix} = \begin{bmatrix} \underline{Y} \\ \overline{Y} \end{bmatrix} \Leftrightarrow \begin{bmatrix} B_{\geq 0} & C \\ C_{\geq 0} & B \end{bmatrix} \begin{bmatrix} \underline{X} \\ -\overline{X} \end{bmatrix} = \begin{bmatrix} \underline{Y} \\ -\overline{Y} \end{bmatrix}$$

5.1.5 Theorem

The vector X is the interval answer of the system $X = S^{-1}X$. If for any arbitrary vector Y, the matrix S^{-1} is non-negative, that is

$$\left(S^{-1}\right)_{ij} \geq 0, \quad 1 \leq i, j \leq 2n$$

Proof: we have

$$\begin{bmatrix} D & E \\ E & D \end{bmatrix} \begin{bmatrix} \underline{Y} \\ -\overline{Y} \end{bmatrix} = \begin{bmatrix} \underline{X} \\ -\overline{X} \end{bmatrix}$$

So, if we put $t_{ij} := \left(S^{-1}\right)_{ij}$, we have

$$\underline{x}_i = \sum_{j=1}^{n} t_{ij}\underline{y}_j - \sum_{j=1}^{n} t_{i,n+j}\overline{y}_j \tag{5.12}$$

$$-\overline{x}_i = \sum_{j=1}^{n} t_{i+n,j}\underline{y}_j - \sum_{j=1}^{n} t_{i+n,n+j}\overline{y}_j, \quad i = 1,\ldots,n \tag{5.13}$$

the equation (5.13) can be written as follows:

$$\overline{x}_i = -\sum_{j=1}^{n} t_{i,n+j}\underline{y}_j + \sum_{j=1}^{n} t_{ij}\overline{y}_j \tag{5.14}$$

From the subtraction of (5.14) from (5.12), we have:

$$\bar{x}_i - \underline{x}_i = \sum_{j=1}^{n}\left(t_{ij} + t_{i+n,j}\right)\left(\bar{y}_j - \underline{y}_j\right)$$

Because all the components of vector Y are intervals and for $1 \le i, j \le 2n$, $t_{ij} \ge 0$, then:

$$\bar{x}_i - \underline{x}_i \ge 0, \quad i = 1,\dots,n$$

As a result, it can be said that all components of vector X are intervals.

It should be noted that it may be that the vector X is an interval but S^{-1} does not be non-negative.

5.1.6 Example

Consider the following tabular data (Table 5.1).

Given that for every x and every j, $x^j \ge 0$, so for system (5.8), the matrix B will be as follows:

$$B = \begin{bmatrix} 1 & 1 & 1 & 1 & 1 & 1 \\ 1 & 2 & 2^2 & 2^3 & 2^4 & 2^5 \\ 1 & 3 & 3^2 & 3^3 & 3^4 & 3^5 \\ 1 & 5 & 5^2 & 5^3 & 5^4 & 5^5 \\ 1 & 6 & 6^2 & 6^3 & 6^4 & 6^5 \\ 1 & 8 & 8^2 & 8^3 & 8^4 & 8^5 \end{bmatrix}$$

and $C = 0$. After solving the system (5.8), $\underline{a}$ and $\bar{a}$ are obtained as follows:

$$\underline{a} = \begin{bmatrix} 2.7143 \\ -5.0881 \\ 4.7627 \\ -1.6010 \\ 0.2230 \\ -0.0109 \end{bmatrix}, \quad \bar{a} = \begin{bmatrix} 3.8571 \\ 2.2726 \\ -1.2395 \\ 0.0808 \\ 0.0323 \\ -0.0034 \end{bmatrix}$$

TABLE 5.1

Data for Example (5.1.6)

x_i	1	2	3	5	6	8
$\left[\underline{f}_i, \bar{f}_i\right]$	[1,5]	[2,4.5]	[2.5,3.5]	[1.5,4]	[2,6]	[3,6]

According to the equation (5.6), we will have:

$$\underline{p}(x) = 2.7143 - 5.0881x + 4.7627x^2 - 1.6010x^3 + 0.2230x^4 - 0.0109x^5$$

$$\overline{p}(x) = 3.8571 + 2.2726x - 1.2395x^2 + 0.0808x^3 + 0.0323x^4 - 0.0034x^5$$

It is obvious that:

$$\left[\underline{p}(x_i), \overline{p}(x_i)\right] = \left[\underline{f_i}, \overline{f_i}\right], \quad i = 0,\dots,5$$

It is clear that $\underline{a}$ and $\overline{a}$ do not form intervals for $i = 0,\dots,5$, but according to the Figure (5.2), $\underline{p}(x)$ and $\overline{p}(x)$ form intervals in the interpolation domain. Note that all calculations are rounded to four decimal places.

5.1.7 Example

Consider the following tabular data (Table 5.2)
we have:

$$
\begin{bmatrix}
1 & 0 & 9 & 0 & 0 & -3 & 0 & -27 \\
1 & 0 & 4 & 0 & 0 & -2 & 0 & -8 \\
1 & 1 & 1 & 1 & 0 & 0 & 0 & 0 \\
1 & 4 & 16 & 64 & 0 & 0 & 0 & 0 \\
0 & -3 & 0 & -27 & 1 & 0 & 9 & 0 \\
0 & -2 & 0 & -8 & 1 & 0 & 4 & 0 \\
0 & 0 & 0 & 0 & 1 & 1 & 1 & 1 \\
0 & 0 & 0 & 0 & 1 & 4 & 16 & 64
\end{bmatrix}
\begin{bmatrix}
\underline{a_0} \\
\underline{a_1} \\
\underline{a_2} \\
\underline{a_3} \\
\overline{a_0} \\
\overline{a_1} \\
\overline{a_2} \\
\overline{a_3}
\end{bmatrix}
=
\begin{bmatrix}
0.5 \\
1 \\
0.5 \\
3.5 \\
1.5 \\
4 \\
3 \\
4
\end{bmatrix}
$$

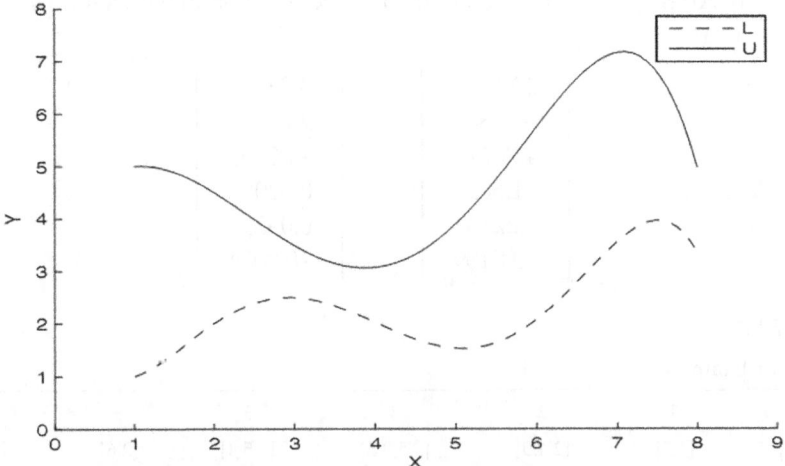

FIGURE 5.2 Interval-valued function for example (5.1.6).

TABLE 5.2

Data for Example (5.1.7)

x_i	−3	−2	1	4
$\left[\underline{f_i}, \overline{f_i}\right]$	[0.5,1.5]	[1,4]	[0.5,3]	[3.5,4]

By solving the above-mentioned system, we will have

$$\underline{a} = \begin{bmatrix} 4.619 \\ -6.3948 \\ 2.5248 \\ -0.249 \end{bmatrix}, \quad \overline{a} = \begin{bmatrix} 0.119 \\ 5.1885 \\ -2.7252 \\ 0.4177 \end{bmatrix}$$

Thus,

$$L:\underline{p}(x) = \begin{cases} 4.619 + 5.1885x + 2.5248x^2 + 0.4177x^3, & x < 0 \\ 4.619 - 6.3948x + 2.5248x^2 - 0.249x^3, & x \geq 0 \end{cases}$$

$$U:\overline{p}(x) = \begin{cases} 0.119 - 6.3948x - 2.7252x^2 - 0.249x^3, & x < 0 \\ 0.119 + 5.1885x - 2.7252x^2 + 0.4177x^3, & x \geq 0 \end{cases}$$

According to the Figure (5.3), it is clear that L and U do not form intervals in a part of the interpolation domain, but satisfy the interpolation condition.

Now, by adding $\left(-1.5,[0.7,2.5]\right), \left(0,[0.1,2]\right)$ and $\left(2.5,[2,3.5]\right)$ to the table, we have:

$$\underline{a} = \begin{bmatrix} 0.1 \\ -34.5418 \\ 90.2276 \\ -87.0004 \\ 39.5157 \\ -8.4973 \\ 0.6961 \end{bmatrix}, \quad \overline{a} = \begin{bmatrix} 2 \\ 35.8338 \\ -89.6648 \\ 86.4207 \\ -39.4365 \\ 8.5556 \\ -0.7087 \end{bmatrix}$$

and as a result,

$$L:\underline{p}(x) =$$

$$\begin{cases} 0.1 + 35.8338x + 90.2276x^2 + 86.4207x^3 + 39.5157x^4 + 8.5556x^5 + 0.6961x^6, & x < 0 \\ 0.1 - 34.5418x + 90.2276x^2 - 87.0004x^3 + 39.5157x^4 - 8.4973x^5 + 0.6961x^6, & x \geq 0 \end{cases}$$

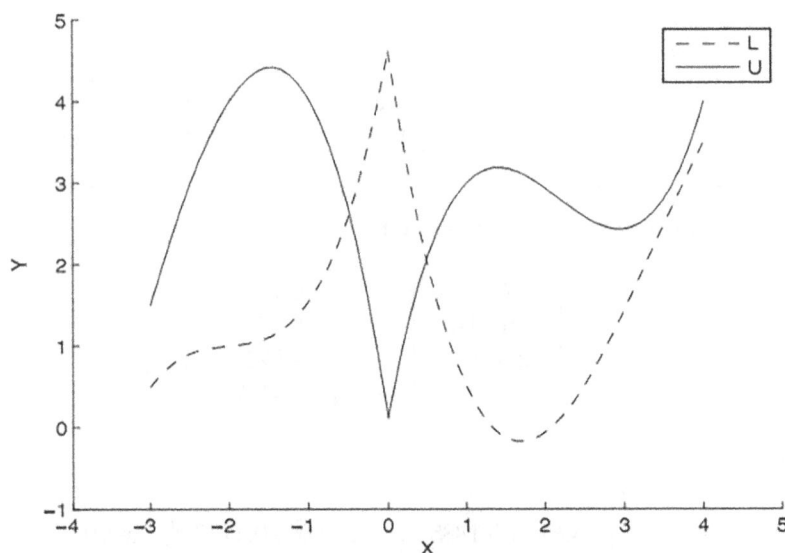

FIGURE 5.3 Interval-valued function for example (5.1.7).

$U : \bar{p}(x) =$

$$\begin{cases} 2 - 34.5418x + 89.6648x^2 - 87.0004x^3 - 39.4365x^4 - 8.4973x^5 - 0.7087x^6, & x < 0 \\ 2 + 35.8338x + 89.6648x^2 + 86.4207x^3 - 39.4365x^4 + 8.5556x^5 - 0.7087x^6, & x \geq 0 \end{cases}$$

According to the Figure (5.4), U and L form intervals in the interpolation domain. Thus, as the number of interpolation points increases, the degree of interpolation polynomials increases and, as a result, the interpolation becomes more accurate. Now, if a more accurate interpolator such as spline is used, the approximation will be much better.

5.1.8 Theorem-Interval Interpolating Polynomial Error

Suppose that

$$\left(x_i, \left[\underline{f_i}, \bar{f_i} \right] \right), \quad i = 0, \dots, n$$

are the interpolation points of the interval function $\left[\underline{f}, \bar{f} \right]$ and $\left[\underline{p}(x), \bar{p}(x) \right]$ are its interval interpolating polynomial. Also assume that the derivatives of functions $\underline{f}$ and $\bar{f}$ are present and continuous up to the $(n+1)$ th order in the domain of the definition of functions $\underline{f}$ and $\bar{f}$. In this case, for every $\bar{x}$, there are δ and ξ numbers of the interval $I[x_0, \dots, x_n]$ such that

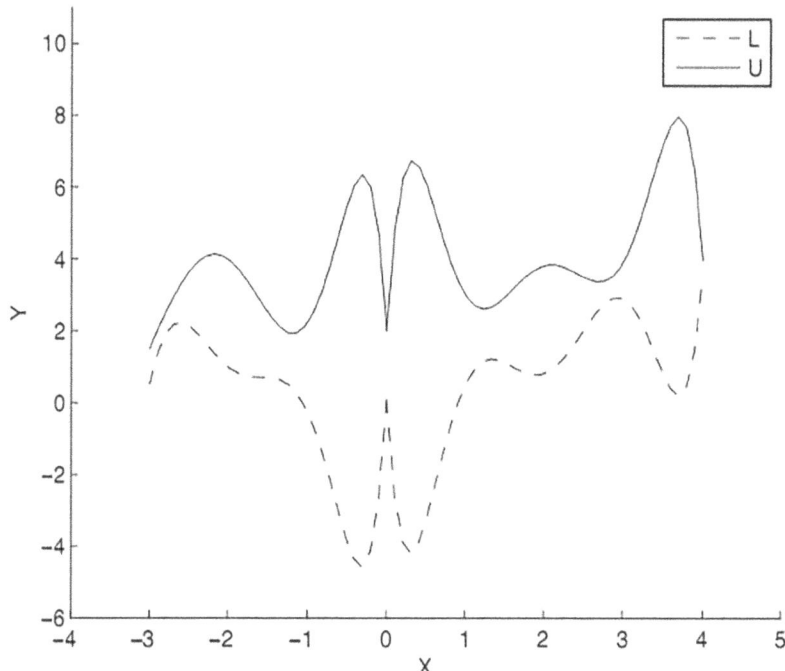

FIGURE 5.4 Interval-valued function for example (5.1.7).

$$\underline{R}(\bar{x}) = \underline{f}(\bar{x}) - \underline{p}(\bar{x}) = \frac{w(\bar{x}) f^{(n+1)}(\delta)}{(n+1)!}$$

$$\overline{R}(\bar{x}) = \overline{f}(\bar{x}) - \overline{p}(\bar{x}) = \frac{w(\bar{x}) \overline{f}^{(n+1)}(\xi)}{(n+1)!}$$

where

$$w(x) = \prod_{i=0}^{n} (x - x_i)$$

Proof: For $i = 0, \ldots, n$, $\bar{x} \neq x_i$ because otherwise, there is nothing left to prove. According to the equations (5.6) and (5.7), we have:

$$\underline{p}(x) = \sum_{\phi_i(x) \geq 0} \underline{a}_i \phi_i(x) + \sum_{\phi_i(x) < 0} \overline{a}_i \phi_i(x)$$

$$\overline{p}(x) = \sum_{\phi_i(x) \geq 0} \overline{a}_i \phi_i(x) + \sum_{\phi_i(x) < 0} \underline{a}_i \phi_i(x)$$

Given that we have $p(x_i) = f_i$ for $i = 0,\ldots,n$, then we can say

$$\underline{R}(x) = \underline{K}\prod_{i=0}^{n}(x - x_i)$$

We choose $\underline{K}$ such that $\overline{x}$ is also the root of the function

$$F(x) = \underline{f}(x) - \underline{p}(x) - \underline{K}w(x) = 0$$

Therefore, it is obvious that the equation $F(x) = 0$ has $n + 2$ roots which are on the interval $I[x_0,\ldots,x_n,\overline{x}]$. According to the mean value theorem, the equation $F'(x) = 0$ has $n + 1$ roots and likewise, the equation $F^{n+1}(x) = 0$ has one root. Suppose that δ is the desired root, that is,

$$\exists\delta\left(\delta \in I[x_0,\ldots,x_n,\overline{x}] \,\&\, F^{n+1}(\delta) = 0\right)$$

So,

$$F^{n+1}(\delta) = \underline{f}^{n+1}(\delta) - \underline{K}(n+1)! = 0$$

And thus,

$$\underline{K} = \frac{\underline{f}^{n+1}(\delta)}{(n+1)!}$$

And so it can be written

$$\overline{K} = \frac{\overline{f}^{n+1}(\xi)}{(n+1)!}$$

If we have

$$\forall x\left(\underline{f}^{n+1}(x) \le \overline{f}^{n+1}(x) \Rightarrow \underline{K} \le \overline{K}\right)$$

In this case, $\underline{R}(x) \le \overline{R}(x)$, and consequently, the error term itself, will be an interval polynomial, but this is not always the case because if $\underline{p}(x)$ is not a good approximation for the points $(x_i, \underline{f_i})$, $i = 0,\ldots,n$ and $\overline{p}(x)$ is a good approximation for the points $(x_i, \overline{f_i})$, $i = 0,\ldots,n$, we may have $\underline{R}(x) \ge \overline{R}(x)$.

5.1.9 Interval Lagrange Interpolation

Now in this section, we examine that how the Lagrange interpolator is obtained using the interval data (x_i s are real numbers and f_i s are intervals).

TABLE 5.3

Interval Data

x_i	x_1	x_2	$\cdots$	x_n
f_i	$\left[\underline{f_1}, \overline{f_1}\right]$	$\left[\underline{f_2}, \overline{f_2}\right]$	$\cdots$	$\left[\underline{f_n}, \overline{f_n}\right]$

Consider the following interval data table (Table 5.3).
According to the Lagrange interpolator polynomials, we have:

$$P(x) = \sum_i f_i L_i(x)$$

$$= \sum_i L_i(x)\left[\underline{f_i}, \overline{f_i}\right]$$

$$= \sum_{L_i(x)<0} L_i(x)\left[\underline{f_i}, \overline{f_i}\right] + \sum_{L_i(x)\geq 0} L_i(x)\left[\underline{f_i}, \overline{f_i}\right]$$

$$= \sum_{L_i(x)<0} \left[L_i(x)\overline{f_i}, L_i(x)\underline{f_i}\right] + \sum_{L_i(x)\geq 0} \left[L_i(x)\underline{f_i}, L_i(x)\overline{f_i}\right]$$

$$= \left[\sum_{L_i(x)<0} L_i(x)\overline{f_i}, \sum_{L_i(x)<0} L_i(x)\underline{f_i}\right] + \left[\sum_{L_i(x)\geq 0} L_i(x)\underline{f_i}, \sum_{L_i(x)\geq 0} L_i(x)\overline{f_i}\right]$$

$$= [L, U]$$

where

$$U = \sum_{L_i(x)<0} L_i(x)\underline{f_i}, \sum_{L_i(x)\geq 0} L_i(x)\overline{f_i}$$

$$L = \sum_{L_i(x)<0} L_i(x)\overline{f_i}, \sum_{L_i(x)\geq 0} L_i(x)\underline{f_i}$$

6

Interpolation from the Linear Algebra Point of View

6.1 Introduction

In this section, we study interpolation from another perspective. As we have already seen, each of the interpolation functions is examined in its own vector space, each of which has at least one basis. If we consider the nature of the interpolation problem, we can say that an interpolation function is obtained by quantifying and forming a system of linear equations. Quantifying means calculating the value of the function at different points x_i and obtaining different values of $f_i \in \mathbb{R}$, so this quantifying operation can be considered as a linear functional such as L so that for the interpolation function p,

$$L_i(p) = p(x_i) = f_i, \quad i = 0, 1, \dots, m \tag{6.1}$$

So according to these explanations, the interpolation problem requires another space, such as functional space or dual space. Obviously, this space also has a dual basis, that is, a basis consisting of the linear functionals L_i. For further explanation, we present the following theorems without proof.

6.1.1 Remark

Suppose that set $\{g_0, \dots, g_n\}$ is a basis for X and $\{L_0, \dots, L_n\}$ is also a basis for X^* space, then

$$\left| L_i(g_j) \right|_{i,j=0}^{n} \neq 0$$

6.1.2 Remark

Suppose $B = \{g_0, \dots, g_n\}$ and $B' = \{L_0, \dots, L_n\}$ are two arbitrary sets of functions and functionals and also suppose $\left| L_i(g_j) \right|_{i,j=0}^{n} \neq 0$. Then, if one of

DOI: 10.1201/9781003218173-6

these two sets is a linearly independent model, so the other has the same properties.

6.1.3 Remark

Suppose that X is a $n+1$ dimensional vector space and $B = \{L_0,\ldots,L_n\}$ is a set on space X^*, then, a necessary and sufficient condition for the interpolation to have a unique answer is that set B be linearly independent.

6.1.4 Corollary

If for every i, $f_i = 0$, then the only answer of interpolation problem is $p \equiv 0$. Now, we define different types of interpolators from the perspective of linear algebra.

6.2 Lagrange Interpolation

Suppose $X = \prod_n (x)$, then

$$\forall p \big(p \in X \Rightarrow L_i(p) = p(x_i), \quad i = 0,1,\ldots,n \big)$$

6.3 Taylor's Interpolation

Suppose $X = \prod_n (x)$, then

$$\forall p \big(p \in X \Rightarrow L_i(p) = p^{(i)}(x_0), \quad i = 0,1,\ldots,n \big)$$

6.4 Abelian Interpolation

Suppose $X = \prod_n (x)$, then

$$\forall p \big(p \in X \Rightarrow L_i(p) = p^{(i)}(x_i), \quad i = 0,1,\ldots,n \big)$$

6.5 Lidestone's Interpolation

Suppose $X = \prod_n (x)$, then

$$\forall p \left(p \in X \Rightarrow L_{2i}(p) = p^{(2i)}(x_0) \ \& \ L_{2i+1}(p) = p^{(2i)}(x_1), \quad i = 0,1,...,n \right)$$

6.6 Simple Hermite Interpolation

Suppose $X = \prod_{2n-1} (x)$, then

$$\forall p \left(p \in X \Rightarrow L_{2i-1}(p) = p(x_i) \ \& \ L_{2i}(p) = p'(x_i), \quad i = 1,...,n \right)$$

6.7 Complete Hermite Interpolation

Suppose $X = \prod_N (x)$, then for every $p \in X$,

$$
\begin{cases}
L_i(p) = p^{(i)}(x_0), & i = 0,1,...,n_0 \\
L_{n_0+i+1}(p) = p^{(i)}(x_1), & i = 0,1,...,n_1 \\
L_{n_0+n_1+2+i}(p) = p^{(i)}(x_2), & i = 0,1,...,n_2 \\
\quad\vdots \\
L_{\sum_{i=0}^{n-1} n_i+n+i}(p) = p^{(i)}(x_n), & i = 0,1,...,n_n
\end{cases}
$$

where $N + 1 = \sum_{i=0}^{n-1} n_i + n + i$.

6.8 Fourier Interpolation

Suppose

$$X = LS\{1, \sin x, \cos x, ..., \sin nx, \cos nx\}$$

(LS^1 means the linear space generated by the set). Then, for every $p \in X$,

$$L_{2i}(p) = \int_{-\pi}^{\pi} p(x)\cos ix\,dx, \quad i = 0,1,\ldots,n$$

$$L_{2i}(p) = \int_{-\pi}^{\pi} p(x)\sin ix\,dx, \quad i = 1,\ldots,n$$

6.8.1 Problem

Is it possible to find a parabola such as $p(x)$ for which the values $p(0)$, $p''(0)$ and $p'''(0)$ are already known?

Solution: Assume the general form of parabola is as $p(x) = a_0 + a_1x + a_2x^2$. We have:

$$L_0(p) = p(0) = a_0$$
$$L_1(p) = p''(0) = 2a_2$$
$$L_2(p) = p'''(0) = 0$$

Now, if we consider the parabola's basis as $\{1, x, x^2\}$, then we have

$$\left| L_i(g_j) \right|^2_{i,j=0} = \begin{vmatrix} 1 & 0 & 0 \\ 0 & 0 & 2 \\ 0 & 0 & 0 \end{vmatrix} = 0$$

Therefore, with the above information, such a parabola cannot be made.

6.8.2 Problem

Suppose

$$p_0(x,y) = 1, \quad p_1(x,y) = x, \quad p_2(x,y) = y$$
$$p_3(x,y) = x^2, \quad p_4(x,y) = xy, \quad p_5(x,y) = y^2,\ldots$$

Show that by having n distinct points (x_i, y_i), $i = 0,\ldots,n-1$, it is not always possible to write a linear combination of $p_0, p_1,\ldots,p_{n-1}$ that have predetermined values at these points.

Solution: We consider the basis as $\{1, x, y, x^2, xy, y^2,\ldots\}$, therefore, we have:

$$\left| L_i(g_i) \right| = \begin{vmatrix} 1 & x_0 & y_0 & x_0^2 & x_0y_0 & y_0^2 & \cdots \\ 1 & x_1 & y_1 & x_1^2 & x_1y_1 & y_1^2 & \cdots \\ \vdots & \vdots & \vdots & \vdots & \vdots & \vdots & \vdots \\ 1 & x_{n-1} & y_{n-1} & x_{n-1}^2 & x_{n-1}y_{n-1} & y_{n-1}^2 & \cdots \end{vmatrix}$$

Now, if we consider the distinct points (x_i, y_i) such that for every i, $x_i = 1$, then, in the above determinants, several columns are equal to one, and this results in $|L_i(g_i)| = 0$. So, it is not possible to write linear combinations of p_i s that have predetermined value at these points.

6.8.3 Problem

Suppose X is the space of polynomials of at most degree n over $[0,1]$ and $0 < x_0 < x_1 < \cdots < x_n < 1$, prove that if

$$L_i : X \to R$$

$$\forall f \left(f \in X \Rightarrow L_i(f) = \int_0^{x_i} f(t)dt \right)$$

then, $\{L_0, L_1, \ldots, L_n\}$ is linearly independent.

Solution: If we consider $\{1, x, \ldots, x^n\}$ as a basis for X such that $x^i = g_i$, $i = 0, \ldots, n$, then

$$L_i = (g_i) = \int_0^{x_i} g_i \, dx$$

Therefore,

$$|L_i(g_i)| = \begin{vmatrix} x_0 & \dfrac{1}{2}x_0^2 & \dfrac{1}{3}x_0^3 & \cdots & \dfrac{1}{n+1}x_0^{n+1} \\ x_1 & \dfrac{1}{2}x_1^2 & \dfrac{1}{3}x_1^3 & \cdots & \dfrac{1}{n+1}x_1^{n+1} \\ \vdots & \vdots & \vdots & \vdots & \vdots \\ x_n & \dfrac{1}{2}x_n^2 & \dfrac{1}{3}x_n^3 & \cdots & \dfrac{1}{n+1}x_n^{n+1} \end{vmatrix}$$

If we factor the coefficients, we have:

$$|L_i(g_i)| = \dfrac{1}{2 \times 3 \times \cdots \times n+1} \begin{vmatrix} x_0 & x_0^2 & x_0^3 & \cdots & x_0^{n+1} \\ x_1 & x_1^2 & x_1^3 & \cdots & x_1^{n+1} \\ \vdots & \vdots & \vdots & \vdots & \vdots \\ x_n & x_n^2 & x_n^3 & \cdots & x_n^{n+1} \end{vmatrix}$$

Since the obtained determinant can be transformed to the Vandermonde determinant, then

$$|L_i(g_i)| \neq 0$$

Then, $\{L_0, L_1, \ldots, L_n\}$ is linearly independent.

6.8.4 Problem

If

$$a_0 + a_1 \cos x + a_2 \cos 2x + a_3 \cos 3x + \cdots + a_n \cos nx$$

is zero at $n+1$ points of $0 \le x_0 < x_1 < \cdots < x_n < \pi$, then it is zero at all the point on the interval $[0, \pi]$.

Solution: To show that the above relation is zero at any point of the interval $[0, \pi]$, it suffices to show that all a_i's are zero. According to the assumption of the problem,

$$\begin{cases} a_0 + a_1 \cos x_0 + a_2 \cos 2x_0 + \cdots + a_n \cos nx_0 = 0 \\ a_0 + a_1 \cos x_1 + a_2 \cos 2x_1 + \cdots + a_n \cos nx_1 = 0 \\ \vdots \\ a_0 + a_1 \cos x_n + a_2 \cos 2x_n + \cdots + a_n \cos nx_n = 0 \end{cases}$$

The above system can be summarized as follows:

$$C = \begin{bmatrix} 1 & \cos x_0 & \cdots & \cos nx_0 \\ 1 & \cos x_1 & \cdots & \cos nx_1 \\ \vdots & \vdots & \vdots & \vdots \\ 1 & \cos x_n & \cdots & \cos nx_n \end{bmatrix}, \quad A = \begin{bmatrix} a_0 \\ a_1 \\ \vdots \\ a_n \end{bmatrix}$$

Now, we have: $CA = 0$.

If we want the system to have a unique answer as $A = 0$, it is enough to show that $|C| \neq 0$ and this is proved by induction on the order of the matrix.

Initial case: If $n = 1$, its determinant is obviously non-zero.

Induction hypothesis: Suppose the determinant of each matrix as C of the order $n = k - 1$ is non-zero.

Induction step: We prove that the determinant of each matrix as C of the order $n = k$ is also non-zero.

$$|C| = \begin{vmatrix} 1 & \cos x_0 & \cdots & \cos kx_0 \\ 1 & \cos x_1 & \cdots & \cos kx_1 \\ \vdots & \vdots & \vdots & \vdots \\ 1 & \cos x_n & \cdots & \cos kx_n \end{vmatrix}$$

If we expand the determinant on the last column, we will have:

$$|C| = \cos kx_0 \cdot |V_0| + \cos kx_1 \cdot |V_1| + \cdots + \cos kx_k \cdot |V_k|$$

Where $|V_i|$ is the determinant obtained by removing the last column and $(i+1)$-th row. As V_i is $(k-1) \times (k-1)$, then according to the induction hypothesis, $|V_i| \neq 0$, $i = 0, \dots, n$ and since the points are distinct, then, $\cos jx_j$ s are not equal to zero all together and that means $|C| \neq 0$.

6.8.5 Problem

If

$$b_1 \sin x + b_2 \sin 2x + \cdots + b_{n-1} \sin(n-1)x + b_n \sin nx$$

at n points of $0 < x_1 < x_2 < \cdots < x_n < \pi$ is equal to zero, then it will be equal to zero at all the points on the interval $[0, \pi]$.

Solution: As with problem (6.8.4), we have:

$$S = \begin{bmatrix} \sin x_1 & \sin x_1 & \cdots & \sin nx_1 \\ \sin x_2 & \sin 2x_2 & \cdots & \sin nx_2 \\ \vdots & \vdots & \vdots & \vdots \\ \sin x_n & \sin 2x_n & \cdots & \sin nx_n \end{bmatrix}, \quad B = \begin{bmatrix} b_1 \\ b_2 \\ \vdots \\ b_n \end{bmatrix}$$

where $SB = 0$. For all values to be zero, we should have $|S| \neq 0$ and this can be proved by induction on the order of the matrix.

Initial case: if $n = 2$, then

$$\begin{vmatrix} \sin x_1 & \sin 2x_1 \\ \sin x_2 & \sin 2x_2 \end{vmatrix} = \sin x_1 \sin 2x_2 - \sin 2x_1 \sin x_2$$

$$= \sin x_1 \cdot 2 \sin x_2 \cos x_2 - \sin x_2 \cdot 2 \sin x_1 \cos x_1$$

$$= 2 \sin x_1 \sin x_2 (\cos x_2 - \cos x_1)$$

$$\neq 0$$

The above term will be zero when $x_1 = x_2 = k\pi$, $k = 0, 1, \dots$ or $\cos x_1 = \cos x_2$ and this is impossible.

Induction hypothesis: We suppose that the determinant of each matrix as S of the order $n = k - 1$ is non-zero.

Induction step: We prove that the determinant of each matrix as S of the order $n = k$ is also non-zero.

$$|S| = \begin{vmatrix} \sin x_1 & \sin x_1 & \cdots & \sin nx_1 \\ \sin x_2 & \sin 2x_2 & \cdots & \sin nx_2 \\ \vdots & \vdots & \vdots & \vdots \\ \sin x_n & \sin 2x_n & \cdots & \sin nx_n \end{vmatrix}$$

If we expand the determinant on the last column,

$$|S| = \sin kx_1 \cdot |V_1| + \sin kx_2 \cdot |V_2| + \cdots + \sin kx_k |V_k|$$

Since V_i are $(k-1)\times(k-1)$, then according to the induction assumption, $|V_i| \neq 0$ and also, as the points are distinct, then $\sin jx_j$ are not equal to zero all together, and therefore, $|S| \neq 0$.

6.8.6 Problem

Show that Lidestone's interpolation is unique.

   Solution: if in the equation $L_i(p) = w_i$, $i = 0,\dots,2n+1$, we consider all w_i to be equal to zero, then we have

$$L_i(p) = 0, \quad i = 0,\dots,2n+1 \tag{6.2}$$

Now, suppose that $p \in X$, then given that $g_0,\dots,g_{2n+1}$ is the basis of $X = \Pi_{2n+1}$, we have:

$$p = a_0 g_0 + a_1 g_1 + \cdots + a_n g_n + \cdots + a_{2n+1} g_{2n+1}$$

That means p is the answer, therefore according to the Lidestone's interpolation, $p^{(i)}(x_0) = 0$, then x_0 is the repeated root of order $n+1$ and so, $p^{(i)}(x_1) = 0$, that is, x_1 is also the root of $n+1$ order. Hence, it can be said that:

$$p(x) = A(x)\left[(x - x_0)^{n+1} \cdot (x - x_1)^{n+1}\right]$$

The phrase in bracket is of degree $2n+2$ and $p \in \Pi_{2n+1}$. So $A(x) \equiv 0$ and this means the answer is unique.

6.8.7 Problem

If $R(x) = \dfrac{A + Bx}{1 + Cx}$, can problem $R(0) = f(0)$, $R'(0) = f'(0)$, $R''(0) = f''(0)$ be always solved?

   Solution:

$$R(x) = \frac{A + Bx}{1 + Cx}, \quad R(0) = A$$

$$R'(x) = \frac{B - CA}{(1 + Cx)^2}, \quad R'(0) = B - CA$$

$$R''(0) = -\frac{2C(b - CA)}{(1 + Cx)^2}, \quad R''(0) = -2C(B - CA)$$

If we write the McLaurin expansion of $R(x)$, we have

$$R(x) = R(0) + xR'(0) + \frac{x^2}{2!}R''(0)$$

$$= A + (B - CA)x - C(B - CA)x^2$$

On the other hand, as $R(x_i) = f(x_i)$, $i = 1, 2, 3$, then the following system is obtained:

$$A + (B - CA)x_1 - C(B - CA)x_1^2 = f_1 \qquad (6.3)$$

$$A + (B - CA)x_2 - C(B - CA)x_2^2 = f_2 \qquad (6.4)$$

$$A + (B - CA)x_3 - C(B - CA)x_3^2 = f_3 \qquad (6.5)$$

If we consider pairs of the above relations, for example, (6.3) and (6.4), we have

$$(B - CA)\left(x_2 - x_1 - C\left(x_2^2 - x_1^2\right)\right) = f_2 - f_1$$

Then,

$$(B - CA)(x_2 - x_1)(1 - C(x_2 + x_1)) = f_2 - f_1$$

Now if $f_1 = f_2$, then we have $x_1 = x_2$ or $B = CA$, $C(x_2 + x_1) = 1$. So, if the points have equal widths, then one of the above states must be established for the system to have an answer, and this is not always the case. Therefore, it cannot be said that the problem can always be solved under those conditions.

7

Newton-Cotes Quadrature

7.1 Newton-Cotes Quadrature

These methods are generally of two types.

1. Closed methods
2. Open methods

Briefly, it can be said that closed methods are methods that use the initial and final points of the integration interval in the quadrature rule, and open methods are methods where at least one of the initial or final points is not used. First, we examine closed methods.

One of the defining factors of these methods is the number of integration points that they use, for example, two points, three points, and so on. It will be proved that the application of eight-point and above methods is not cost-effective and the propagation of computational and rounding error is occurred. Each of the above methods is based on the choice of interpolating polynomials that are used instead of f, because in the quadrature rule $\left(I(f)\right)$, interpolating polynomials are used instead of the function f. For two reasons, one is that we may not have the rule of the function f and we may only have information about it, and the other is that it is much easier to integrate polynomials rather than other functions. Now if we use a linear interpolator, the method is a two-point method, and if one of the points is not used (open Newton-Cotes), the method is a one-point method. If we use a parabola as an interpolator, the method is a three-point method and so on. Obviously, by using the interpolator instead of the function f and integrating it as an approximation, $I(f)$) will have an error, because the interpolator has an error. That is, if $p_n(x)$ is the interpolator of $f(x)$ on the interval $[a,b]$ with an error of $R_n(x)$, we have:

$$f(x) = p(x) + R_n(x)$$

DOI: 10.1201/9781003218173-7

So,

$$\int_a^b f(x)dx = \int_a^b p_n(x)dx + \int_a^b R_n(x)dx$$

That is,

$$I(f) = Q(f) + E(f), \quad E(f) = \int_a^b R_n(x)dx$$

where $E(f)$ is the integration error.
   Now suppose that

$$p_n(x) = \sum_{i=0}^n L_i(x)f_i, \quad L_i(x) = \prod_{j=0, j\neq i}^n \frac{x-x_j}{x_i-x_j}$$

where $x_j = a + jh$ and $h = \dfrac{b-a}{n}$. By changing the variables as $x = a + th$ where $0 \le t \le n$, it can be easily shown that

$$L_i(x) = \prod_{j=0, j\neq i}^n \frac{t-j}{i-j}$$

So,

$$\int_a^b p_n(x)dx = h\sum_{i=0}^n \alpha_i f_i = \sum_{i=0}^n w_i f_i \tag{7.1}$$

where

$$\alpha_i = \int_0^n \varphi_i(x)dt \tag{7.2}$$

The following problem shows that the Newton-Cotes integration weights are unique.

7.1.1 Problem

Suppose that $\Delta = \{a = x_0 < x_1 < \cdots < x_n = b\}$ is a fixed and arbitrary partition of the interval $[a,b]$. Show that for any polynomial p where $\deg(p) \le n$, there are unique numbers $\gamma_0, \ldots, \gamma_n$ such that:

$$\sum_{i=0}^n \gamma_i p(x_i) = \int_a^b p(x)dx$$

Solution: Suppose that $p(x) = 1, x, x^2, \ldots, x^n$, in this case we have:

$$p(x) = 1 \quad \Rightarrow \quad \gamma_0 + \cdots + \gamma_n = b - a$$

$$p(x) = x \quad \Rightarrow \quad \gamma_0 x_0 + \cdots + \gamma_n x_n = \frac{b^2 - a^2}{2}$$

$$p(x) = x^2 \quad \Rightarrow \quad \gamma_0 x_0^2 + \cdots + \gamma_n x_n^2 = \frac{b^3 - a^3}{3}$$

$$\vdots$$

$$p(x) = x^n \quad \Rightarrow \quad \gamma_0 x_0^n + \cdots + \gamma_n x_n^n = \frac{b^{n+1} - a^{n+1}}{n+1}$$

The determinant of the multiplication matrix of the above system is a Vandermonde determinant, which is non-zero due to the distinction of x_is.

$$\begin{vmatrix} 1 & 1 & \cdots & 1 \\ x_0 & x_1 & \cdots & x_n \\ x_0^2 & x_1^2 & \cdots & x_n^2 \\ \vdots & \vdots & \vdots & \vdots \\ x_0^n & x_1^n & \cdots & x_n^n \end{vmatrix} \neq 0$$

so, the system will have a unique answer set such as $\gamma_0, \ldots, \gamma_n$.

7.1.2 Problem

Prove that

$$\sum_{i=0}^{n} \alpha_i = n \tag{7.3}$$

$$\alpha_i = \alpha_{n-i} \tag{7.4}$$

Solution: Since the quadrature rule holds for any arbitrary function f, let us assume that $f(x) \equiv 1$, so we have $p(x) \equiv 1$. According to the Newton-Cotes quadrature rule:

$$\int_a^b p(x)\,dx = h \sum_{i=0}^{n} f_i \alpha_i$$

where $h = \dfrac{b-a}{n}$. Thus,

$$\int_a^b dx = h \sum_{i=0}^{n} \alpha_i$$

Therefore,

$$\sum_{i=0}^{n} \alpha_i = n$$

According to the equation (7.3), we have:

$$\alpha_i = \int_0^n \varphi_i(\theta)d\theta, \quad \alpha_{n-i} = \int_0^n \varphi_{n-i}(\theta)d\theta$$

Then,

$$\int_0^n \varphi_i(\theta)d\theta = \int_0^n \prod_{j=0, j\neq i}^{n} \frac{\theta - j}{i - j} d\theta$$

$$= \int_0^n \frac{\theta(\theta - 1)...(\theta - (i-1))(\theta - (i+1))...(\theta - n)}{(i-0)(i-1)...(i-(i-1))(i-(i+1))...(i-n)} d\theta$$

$$= \int_0^n \frac{\theta(\theta - 1)...(\theta - i + 1)(\theta - i - 1)...(\theta - n)}{(-1)^{n-i} \cdot i! \cdot (n-i)!} d\theta$$

Also, by assuming $l = n - i$, we have:

$$\int_0^n \varphi_{n-i}(\theta)d\theta = \int_0^n \varphi_l(\theta)d\theta$$

$$= \int_0^n \frac{\theta(\theta - 1)...(\theta - (l-1))(\theta - (l+1))...(\theta - n)}{(l-0)(l-1)...(l-(l-1))(l-(l+1))...(l-n)} d\theta$$

Now, by assuming $\theta = n - x$, we will finally have:

$$\int_0^n \varphi_{n-i}(\theta)d\theta = \int_n^0 \frac{(n-x)...(-x+i+1)(-x+i-1)...(-x)}{(-1)^i \cdot i! \cdot (n-i)!} d(-x)$$

$$= \int_0^n \frac{(-1)^n (x-n)...(x-(i+1))(x-(i-1))...x}{(-1)^i \cdot i! \cdot (n-i)!} dx$$

By using a change of variables as $x = \theta$, we can conclude

$$\int_0^n \varphi_{n-i}(\theta)d\theta = \int_0^n \frac{\theta(\theta - 1)...(\theta - i + 1)(\theta - i - 1)...(\theta - n)}{(-1)^{n-i} \cdot i! \cdot (n-i)!} d\theta$$

$$= \int_0^n \varphi_i(\theta)d\theta$$

7.1.3 Problem

If we assume

$$c_i^n = \frac{1}{n} \int_0^n \varphi_i(\theta) d\theta \tag{7.5}$$

In this case, show

$$\sum_{i=0}^{n} c_i^n = 1 \tag{7.6}$$

$$c_i^n = c_{n-i}^n \tag{7.7}$$

Solution: To prove the equation (7.6), we write:

$$\int_a^b L_i(x) dx = h \int_0^n \varphi_i(\theta) d\theta$$

$$= h \cdot n \cdot \frac{1}{n} \int_0^n \varphi_i(\theta) d\theta$$

$$= \frac{b-a}{n} \cdot n \left(\frac{1}{n} \int_0^n \varphi_i(\theta) d\theta \right)$$

$$= (b-a) \cdot n \left(\frac{1}{n} \int_0^n \varphi_i(\theta) d\theta \right)$$

$$= (b-a) c_i^n$$

So,

$$\sum_{i=0}^{n} c_i^n = \sum_{i=0}^{n} \frac{1}{b-a} \int_a^b L_i(x) dx$$

$$= \frac{1}{b-a} \int_a^b \left(\sum_{i=0}^{n} L_i(x) \right) dx$$

$$= \frac{1}{b-a} \int_a^b dx$$

$$= 1$$

Also, according to the equation (7.4), we have:

$$\frac{1}{n} \alpha_i = \frac{1}{n} \alpha_{n-i}$$

So according to the equation (7.3):

$$c_i^n = c_{n-i}^n$$

7.2 The Peano's Kernel Error Representation

In this subject, the aim is to obtain a constant coefficient of the Newton-Cotes error formula by providing a simpler and easier method. Suppose that we have $R(p) = 0$ for each polynomial of $p \in \Pi_n$. That is, the quadrature rule is exact for all polynomials of at most degree n. In this case, for each function $f \in c^{n+1}[a,b]$, we have:

$$R(f) = \int_a^b f^{(n+1)}(t) K(t) dt$$

where

$$K(t) := \frac{1}{n!} R_x \left((x-t)_+^n \right), (x-t)_+^n := \begin{cases} (x-t)^n, & x \geq t \\ 0, & x < t \end{cases}$$

and $R_x \left((x-t)_+^n \right)$ indicates the error $(x-t)_+^n$. The function $K(t)$ is usually called operator kernel R or Peano kernel.

If the sign of function $K(t)$ does not change on the integration interval, then we use the following equation to estimate the error:

$$R_n(f) = \begin{cases} \dfrac{R_n(x^{n+1})}{(n+1)!} f^{(n+1)}(\xi), & \text{odd } n \\ \dfrac{R_n(x^{n+2})}{(n+2)!} f^{(n+2)}(\xi), & \text{even } n \end{cases} \qquad \xi \in [a,b]$$

7.2.1 Problem

Specify the Peano kernel for the Simpson's quadrature rule with $n = 2$ instead of $n = 3$ on the interval $[-1,1]$. Does the sign of Peano kernel change on this interval.

Solution:

$$K(t) = \frac{1}{n!} R_x \left((x-t)_+^n \right)$$

$$= \frac{1}{2!} \left(\int_{-1}^1 (x-t)_+^2 \, dx - \frac{1}{3} \left((-1-t)_+^2 + 4(0-t)_+^2 + (1-t)_+^2 \right) \right)$$

$$\Rightarrow K(t) = \begin{cases} \dfrac{1}{2}\left(\dfrac{(1-t)^3}{3} - \dfrac{1}{3}(1-t)^2\right) = -\dfrac{1}{6}(1-t)^2 t & t > 0 \\[4mm] \dfrac{1}{2}\left(\dfrac{(1-t)^3}{3} - \dfrac{1}{3}\left(4t^2 + (1-t)^2\right)\right) = -\dfrac{1}{6}(1+t)^2 t & t \le 0 \end{cases}$$

where $K(t)$ is non-positive on the interval $[0,1]$ and non-negative on the interval $[0,-1]$.

7.3 Romberg's Quadrature Rule

In this section, we describe a way to increase the accuracy of the quadrature. This method works recursively and estimates the integration methods with the accuracy $o(h^{n+1})$ from the integration methods with the accuracy $o(h^n)$. This method operates similar to the divided difference interpolation or Neville's methods, which move in the form of an equilateral triangle. This method is used for all integration rules that are as $\sum_i w_i f_i$ or a linear combination of function values at the integration interval points. One way to obtain such series is to introduce Euler-Macluarin expansion methods. All of the above topics are discussed in detail in [1]. In general, the following recursive relation can be written for the Romberg quadrature rule:

$$T_{m,k} = \frac{1}{4^m - 1}\left(4^m T_{m-1,k+1} - T_{m-1,k}\right), \quad m = 1,2,\dots \quad k = 0,1,\dots$$

With repeated applications of the rule, it can be shown that

$$T_{m,k} = \sum_{l=0}^{m} c_{m,m-l} T_{0,k+l}$$

The process that was mentioned for approximation $I(f)$ is shown in Table 7.1.

7.3.1 Problem

Using the Euler-Macluarin formula, prove that:

$$\sum_{k=0}^{n} k^3 = \left(\frac{n(n+1)}{2}\right)^2$$

TABLE 7.1

Romberg's Quadrature Rule

$\dfrac{1}{6!}f^{(6)}(\xi)$	$T_{0,j}$	$T_{1,j}$	$T_{2,j}$	$\cdots$	$T_{m,j}$
$\dfrac{b-a}{2^{k}}$	$T_{0,0}$				
$\dfrac{b-a}{2^{k+1}}$	$T_{0,1}$	$T_{1,0}$			
$\dfrac{b-a}{2^{k+2}}$	$T_{0,2}$	$T_{1,1}$	$T_{2,0}$		
$\vdots$	$\vdots$				
$\dfrac{b-a}{2^{k+m}}$	$T_{0,m}$	$T_{1,m-1}$	$T_{2,m-2}$	$\cdots$	$T_{m,0}$

Solution: According to Euler-Macluarin formula,

$$\int_0^n g(t)\,dt = \frac{1}{2}g(0)+g(1)+\cdots+g(n-1)+\frac{1}{2}g(n)+$$

$$\sum_{l=1}^m \frac{B_{2l}}{(2l)!}\left(g^{(2l-1)}(0)-g^{(2l-1)}(n)\right)-\frac{B_{2m+2}}{(2m+2)!}n\cdot g^{(2m+2)}(\xi) \quad \xi\in(0,n)$$

Suppose that $g(t)=t^3$,
   therefore

$$g'(t)=3t^2, \quad g''(t)=6t, \quad g^{(3)}(t)=6$$

and if $k>3$, then $g^{(k)}(t)=0$.
   So,

$$g^{(2m+2)}(\xi)=0, \quad g^{(2l-1)}(0)-g^{(2l-1)}(n)=0 \quad l\geq 2$$

Then,

$$\int_0^n t^3\,dt = 0+1^3+2^3+\cdots+(n-1)^3+\frac{n^3}{2}+\frac{B_2}{2}\left(g'(0)-g'(n)\right)$$

$$+\frac{B_4}{4!}\left(g^{(3)}(0)-g^{(3)}(n)\right)$$

As a result, we will have

$$\sum_{k=0}^{n} k^3 = \frac{n^4}{4} + \frac{n^3}{2} + \frac{3n^2}{12}$$

$$= \frac{n^2\left(n^2 + 2n + 1\right)}{4}$$

$$= \left(\frac{n(n+1)}{2}\right)^2$$

7.3.2 Problem

Suppose that $h_0 := b - a$ and $h_1 = \dfrac{h_0}{3}$. Show that extrapolations of $T(h_0)$ and $T(h_1)$ give the rule $\dfrac{3}{8}$.

Solution:

$$T_{00} = \frac{h_0}{2}\left(f(a) + f(b)\right)$$

$$T_{10} = \frac{h_0}{6}\left(f(a) + 2f(a + h_1) + 2f(a + 2h_1) + f(b)\right)$$

$$T_{11} = T_{10} + \frac{T_{10} - T_{00}}{\left(\dfrac{h_0}{h_1}\right)^2 - 1}$$

Thus,

$$T_{11} = \frac{h_0}{6}\left(f(a) + 2f(a + h_1) + 2f(a + 2h_1) + f(b)\right)$$

$$+ \frac{1}{8}\left(\frac{h_0}{6}\left(f(a) + 2f(a + h_1) + 2f(a + 2h_1) + f(b)\right) - \frac{h_0}{2}\left(f(a) + f(b)\right)\right)$$

$$= \frac{h_0}{6}\left(f(a) + 2f(a + h_1) + 2f(a + 2h_1) + \frac{1}{8}f(a)\right)$$

$$+ \frac{h_0}{6}\left(\frac{1}{4}f(a + h_1) + \frac{1}{4}f(a + 2h_1) + \frac{1}{8}f(b) - \frac{3}{8}f(a) - \frac{3}{8}f(b)\right)$$

$$= \frac{h_0}{6}\left(\frac{6}{8}f(a) + \frac{6}{8}f(b) + \frac{9}{4}f(a + h_1) + \frac{9}{4}f(a + 2h_1)\right)$$

$$= \frac{h_0}{3}\left(\frac{3}{8}f(a) + \frac{3}{8}f(b) + \frac{9}{8}f(a + h_1) + \frac{9}{8}f(a + 2h_1)\right)$$

$$= \frac{3}{8}h_1\left(f(a) + 3f(a + h_1) + 3f(a + 2h_1) + f(b)\right)$$

7.3.3 Problem

Consider an integration by a polynomial extrapolation rule, based on a sequence with geometric ratio $h_j = h_0 b^j$, $0 < b < 1$ and $j = 0,1,\ldots,m$. Show that a small error ΔT_j in calculating the trapezoid set $T(h_j)$, $j = 0,\ldots,m$ causes an error ΔT_{mm} in the extrapolated value T_{mm} according to the following equation:

$$\left|\Delta T_{mm}\right| \le c_m\left(b^2\right)\max_{0 \le j \le m}\left|\Delta T_j\right|$$

where $c_m(\theta)$ is defined as follows:

$$c_m(\theta) := \prod_{l=1}^{m}\frac{1+\theta^l}{1-\theta^l}$$

Note that just as when θ tends to 1, $c_m(\theta)$ tends to infinity, so the stability of the extrapolation method is lost, when b tends to 1.

Solution: We know that the asymptotic expansion $T(h)$ is as follows:

$$T(h) = \tau_0 + \tau_1 h^{\gamma_1} + \cdots + \tau_m h^{\gamma_m} + h^{\gamma_{m+1}}\alpha_{m+1}(h)$$

where $0 < \gamma_1 < \cdots < \gamma_{m+1}$ and the coefficients τ_i are independent of h and $\alpha_{m+1}(h)$ is a bounded function, and $\tau_0 = \lim_{h \to 0} T(h)$ is the exact answer to the problem. Now suppose that $\gamma_k = 2k$, then

$$T(h) = \tau_0 + \tau_1 h^2 + \tau_2 h^4 + \cdots + \tau_m h^{2m} + h^{2m+2}\alpha_{m+1}(h)$$

where $\tau_0 = f'(x)$ and $\tau_k = \dfrac{f^{(2k+1)}(x)}{(2k+1)!}$, $1 \le k \le m+1$

On the other hand

$$\left|\Delta T_{mm}\right| = \left|T_{mm} - \tau_0\right| \le M_{m+1}c_m\left(b^2\right)z_0 z_1 \cdots z_m \tag{7.8}$$

where $M_{m+1} = \max_{0 \le j \le m}\left|\alpha_{m+1}(h_j)\right|$ and $z_j = h_j^2$.

Also:

$$\Delta T_j = T_{mm} - \int_a^b f(x)dx = (b-a)h_0^2 h_1^2 \ldots h_m^2 (-1)^m \frac{B_{2m+2}}{(2m+2)!} f^{(2m+2)}(\xi)$$

where $a < \xi < b$ and $f \in c^{2m+2}[a,b]$

As

$$\alpha_{m+1}(h) = \frac{B_{2m+2}}{(2m+2)!}(b-a)f^{(2m+2)}(\xi)$$

So, we will have

$$\Delta T_j = h_0^2 h_1^2 \dots h_m^2 (-1)^m \alpha_{m+1}(h)$$

Thus,

$$\max_{0 \le j \le m} |\Delta T_j| = h_0^2 h_1^2 \cdots h_m^2 M_{m+1} \tag{7.9}$$

It follows from the equations (7.8) and (7.9) that:

$$|\Delta T_{mm}| \le c_m (b^2) \max_{0 \le j \le m} |\Delta T_j|$$

7.3.4 Problem

If we represent $T(f,h)$ as the trapezoid sum with the ratio h for $\int_0^1 f(x)dx$, for $\alpha > 1$, $T(x^\alpha, h)$ has an asymptotic expansion as follows:

$$T(x^\alpha, h) \sim \int_0^1 x^\alpha \, dx + a_1 h^{1+\alpha} + a_2 h^2 + a_4 h^4 + \cdots$$

As a result, they show that each function $f(x)$, which is analytical on the closed circle $|z| \le r, (r > 1)$ in complex plane, follows an asymptotic expansion as follows:

$$T(x^\alpha f(x), h) \sim \int_0^1 x^\alpha f(x)dx + b_1 h^{1+\alpha} + b_2 h^{2+\alpha} + b_3 h^{3+\alpha} + \cdots + c_2 h^2 + c_4 h^4 + c_6 h^6 + \cdots$$

Solution: We write the Taylor expansion of function 1 about point $x = 0$,

$$f(x) = f(0) + xf'(0) + \frac{x^2}{2!}f''(0) + \cdots$$

$$= k_0 + k_1 x + k_2 x^2 + \cdots$$

Thus,

$$T\left(x^{\alpha}f(x),h\right)=T\left(x^{\alpha}\left(k_0+k_1x+k_2x^2+\cdots\right),h\right)$$

$$=T\left(k_0\ x^{\alpha},h\right)+T\left(k_1\ x^{1+\alpha},h\right)+T\left(k_2x^{2+\alpha},h\right)+\cdots$$

$$=\int_0^1 k_0x^{\alpha}\ dx+a_1h^{1+\alpha}+a_2h^2+a_4h^4+\cdots$$

$$+\int_0^1 k_1x^{1+\alpha}\ dx+a_1'h^{2+\alpha}+a_2'h^2+a_4'h^4+\cdots$$

$$+\int_0^1 k_2x^{2+\alpha}\ dx+a_1''h^{3+\alpha}+a_2''h^2+a_4''h^4+\cdots$$

$$=\int_0^1\left(k_0+k_1x+k_2x^2+\cdots\right)x^{\alpha}\ dx$$

$$+a_1h^{1+\alpha}+a_1'h^{2+\alpha}+a_1''h^{3+\alpha}+\cdots$$

$$+\left(a_2+a_2'+a_2''+\cdots\right)h^2+\left(a_4+a_4'+a_4''+\cdots\right)h^4+\cdots$$

$$\sim\int_0^1 x^{\alpha}f(x)dx+b_1h^{1+\alpha}+b_2h^{2+\alpha}+b_3h^{3+\alpha}+\cdots$$

$$+c_2h^2+c_4h^4+c_6h^6+\cdots$$

8

Interval Newton-Cotes Quadrature

8.1 Introduction

According to approximation theory, the problem of integration plays a major role in various fields such as mathematics, physics, statistics, engineering, and social sciences. Because in many applications, at least some system parameters and measurements are represented by interval numbers rather than real numbers, it is important to develop interval integrals and solve them. The concept of interval numbers and arithmetic operations with these numbers has already been discussed. Here, we introduce the integration formulas of Newton-Coates methods for the interval integrals with the Peano's error representation theorem as well as the convergence theorem.

8.2 Some Definitions

We represent an arbitrary interval number by an ordered pair of functions $[\underline{u}, \overline{u}]$, such that $\underline{u} \leq \overline{u}$.

For simplification, we name the set of all intervals by E^1.

8.2.1 Lemma

Let v, w are two intervals and s be a real number:
$u = v$ if and only if $\underline{u} = \underline{v}$ and $\overline{u} = \overline{v}$

$$v + w = \left[\underline{v} + \underline{w}, \overline{v} + \overline{w}\right],$$

$$v - w = \left[\underline{v} - \overline{w}, \overline{v} - \underline{w}\right]$$

$$v.w = \min\left[\underline{v}.\underline{w}, \underline{v}.\,\overline{w}, \overline{v}.\,\underline{w}, \overline{v}.\overline{w}\right]$$

$$= \max\left[\underline{v}.\underline{w}, \underline{v}.\,\overline{w}, \overline{v}.\,\underline{w}, \overline{v}.\overline{w}\right]$$

DOI: 10.1201/9781003218173-8

$$s = [s\underline{v},\ s\bar{v}],\, s \geq 0$$

$$s = [s\bar{v}, s\underline{v},],\, s < 0$$

8.2.2 Definition-Distance between Two Intervals

For arbitrary interval numbers $u = [\underline{u}, \bar{u}]$ and $v = [\underline{v}, \bar{v}]$ the quantity

$$D(u,v) = \max\left\{|\underline{u} - \underline{v}|, |\bar{u} - \bar{v}|\right\}$$

is the distance between u and v.

8.2.3 Definition-Continuity of an Interval Function

A function $f : R^1 \to E^1$ is called an interval valued function. If for arbitrary fixed $t_0 \in R^1$ and $\varepsilon > 0, \delta > 0$ such that

$$|t - t_0| < \delta \to D\big[f(t), f(t_0)\big] < \varepsilon$$

exists, f is said to be continuous.

Throughout this chapter, we also consider interval functions which are defined only over a finite interval $[a,b]$. We now define the integral of an interval function using the Rieman integral concept.

8.2.4 Definition

Let $f : [a,b] \to E^1$. For each partition $P = \{t_0, t_1, \ldots, t_n\}$ of $[a,b]$ and for arbitrary $\xi_i : t_{i-1} \leq \xi_i \leq t_i,\, 1 \leq i \leq n$, let

$$R_p = \sum_{i=1}^{n} f(\xi_i)(t_i - t_{i-1})$$

The definite integral of $f(t)$ over $[a,b]$ is

$$\int_a^b f(t)dt = \lim R_p, \quad \max_{1 \leq i \leq n}|t_i - t_{i-1}| \to 0$$

provided that this limit exists in the metric D.

If the interval function $f(t)$ is continuous in the metric D, its definite integral exists. Furthermore,

$$\left(\overline{\int_a^b f(t)dt}\right) = \int_a^b \overline{f}(t)dt, \tag{8.1}$$

$$\left(\overline{\int_a^b f(t)dt}\right) = \int_a^b \overline{f}(t)dt$$

It should be noted that the interval integral can be also defined using Lebesgue-type approach.

8.3 Newtons-Cotes Method

Let f be an interval function. For any natural number n, the Newtons-Cotes formulas

$$\int_a^b f(x)dx = h\sum_{i=1}^n f_i\alpha_i + E, \quad f_i = f(a+ih), \quad h = \frac{b-a}{n},$$

provide approximate values for $\int_a^b f(x)dx$. The interval form of the equation above is as follows:

$$\int_a^b \underline{f}(x)dx = h\sum_{i=1}^n \alpha_i \underline{f}(x_i) + E\left(\underline{f}\right) \tag{8.2}$$

$$\int_a^b \overline{f}(x)dx = h\sum_{i=1}^n \alpha_i \overline{f}(x_i) + E\left(\overline{f}\right)$$

The weights α_i, $i = 1, \dots, n$, are rational numbers with the property $\sum_{i=1}^n \alpha_i = n$. This follows (8.2) when applied to $\underline{f}(x) = \overline{f}(x) = 1$. It can be shown that the approximation error may be expressed as follows:

$$E\left(\underline{f}\right) = h^{p+1} \cdot K \cdot \underline{f}^{(p)}\left(\underline{\zeta}\right), \quad \underline{\zeta} \in [a,b],$$

$$E\left(\overline{f}\right) = h^{p+1} \cdot K \cdot \overline{f}^{(p)}\left(\overline{\zeta}\right), \quad \overline{\zeta} \in [a,b] \tag{8.3}$$

The values of p and K depend only on n but not on the integrand f.

Note
We assume that the arbitrary derivatives of the function f at different points in its domain are defined as intervals.

For large n, some of the values α_i become negative and the corresponding formulas are unsuitable for numerical purposes, as cancellations tend to occur in computing the sum (8.3).

Let

$$Q\left(\underline{f}\right) = h\sum_{i=1}^{n} \alpha_i \underline{f}(x_i),\qquad(8.4)$$

$$Q\left(\overline{f}\right) = h\sum_{i=1}^{n} \alpha_i \overline{f}(x_i),$$

Thus from (8.2) we have

$$I\left(\underline{f}\right) = \int_a^b \underline{f}(x)dx = Q\left(\underline{f}\right) + E\left(\underline{f}\right)\qquad(8.5)$$

$$I\left(\overline{f}\right) = \int_a^b \overline{f}(x)dx = Q\left(\overline{f}\right) + E\left(\overline{f}\right)$$

8.3.1 Peano's Error Representation

From (8.2), we have

$$E\left(\underline{f}\right) = I\left(\underline{f}\right) - Q\left(\underline{f}\right),\qquad(8.6)$$

$$E\left(\overline{f}\right) = I\left(\overline{f}\right) - Q\left(\overline{f}\right),$$

The integration error $E(f)$ is a linear operator in interval form

$$E\left(\alpha\underline{f} + \underline{g}\right) = \alpha E\left(\underline{f}\right) + E\left(\underline{g}\right),$$

$$E\left(\alpha\overline{f} + \overline{g}\right) = \alpha E\left(\overline{f}\right) + E\left(\overline{g}\right),$$

for $f, g \in \tilde{V}, \alpha \in R$ on some appropriate interval function space $\tilde{V}$.

The following elegant interval integral representation of the error $E(f)$ is a classical result due to the Peano's theorem.

8.3.2 Theorem

Suppose $E(p) = 0$ holds for all $p \in \prod_n$, that is, every polynomial whose degree does not exceed n is integrated exactly. Then for all interval functions in parametric form $\underline{f}, \overline{f} \in C^{n+1}[a,b]$,

$$E\left(\underline{f}\right) = \int_a^b \underline{f}^{n+1}(t) K(t) dt, \qquad (8.7)$$

$$E\left(\overline{f}\right) = \int_a^b \overline{f}^{n+1}(t) K(t) dt,$$

where

$$K(t) := \frac{1}{n!} E_x \left[(x-t)_+^n \right], \quad (x-t)_+^n := \begin{cases} (x-t)^n, & x \geq t \\ 0, & x < t \end{cases}$$

and $E_x \left[(x-t)_+^n \right]$, when the latter is considered as a function in x.

Proof: Consider the Taylor expansion of $f(x)$, $x \in [a,b]$ at a, in parametric form:

$$\underline{f}(x) = \underline{f}(a) + \underline{f}'(a)(x-a) + \cdots + \frac{\underline{f}^{(n)}(a)}{n!}(x-a)^n + \underline{M}_n(x), \qquad (8.8)$$

$$\overline{f}(x) = \overline{f}(a) + \overline{f}'(a)(x-a) + \cdots + \frac{\overline{f}^{(n)}(a)}{n!}(x-a)^n + \overline{M}_n(x),$$

Its reminder term can be expressed in the form

$$\underline{M}_n(x) = \frac{1}{n!} \int_a^x \underline{f}^{n+1}(t)(x-t)^n dt = \frac{1}{n!} \int_a^b \underline{f}^{n+1}(t)(x-t)_+^n dt, \qquad (8.9)$$

$$\overline{M}_n(x) = \frac{1}{n!} \int_a^x \overline{f}^{n+1}(t)(x-t)^n dt = \frac{1}{n!} \int_a^b \overline{f}^{n+1}(t)(x-t)_+^n dt,$$

Applying the linear operator E to (8.8) gives

$$E\left(\underline{f}\right) = E(\underline{M}_n) = \frac{1}{n!} E_x \left(\int_a^b \underline{f}^{n+1}(t)(x-t)_+^n dt \right), \qquad (8.10)$$

$$E\left(\overline{f}\right) = E(\overline{M}_n) = \frac{1}{n!} E_x \left(\int_a^b \overline{f}^{n+1}(t)(x-t)_+^n dt \right)$$

Since $E(p) = 0$ for $p \in \prod_n$. The entire operator E_x commutes with integration and we obtain the desired result (8.7).

8.4 Trapezoidal Integration Rule

According to (8.2), we have:

$$I\left(\underline{f}\right) = h\left[\frac{1}{2}\underline{f}(x_0) + \sum_{i=1}^{n-1}\underline{f}(x_i) + \frac{1}{2}\underline{f}(x_n)\right] + E\left(\underline{f}\right), \tag{8.11}$$

$$I\left(\overline{f}\right) = h\left[\frac{1}{2}\overline{f}(x_0) + \sum_{i=1}^{n-1}\overline{f}(x_i) + \frac{1}{2}\overline{f}(x_n)\right] + E\left(\overline{f}\right).$$

where

$$E\left(\underline{f}\right) = -\frac{h^2}{12}(b-a)\underline{f}^{(2)}\left(\underline{\zeta}\right), \tag{8.12}$$

$$E\left(\overline{f}\right) = -\frac{h^2}{12} = (b-a)\overline{f}^{(2)}\left(\overline{\zeta}\right), \quad \underline{\zeta},\overline{\zeta} \in [x_0, x_n]$$

The following variant of the Trapezoidal sum is already a method of order 3:

$$I\left(\underline{f}\right) = h\left[\frac{5}{12}\underline{f}(x_0) + \frac{13}{12}\sum_{i=2}^{n-2}\underline{f}(x_i) + \frac{13}{12}\underline{f}(x_{n-1})\right] + \frac{5}{12}\underline{f}(x_n), \tag{8.13}$$

$$I\left(\overline{f}\right) = h\left[\frac{5}{12}\overline{f}(x_0) + \frac{13}{12}\sum_{i=2}^{n-2}\overline{f}(x_i) + \frac{13}{12}\overline{f}(x_{n-1})\right] + \frac{5}{12}\underline{f}(x_n).$$

8.5 Simpson Integration Rule

According to (8.1), we have:

$$I\left(\underline{f}\right) = h\left[\frac{1}{3}\underline{f}(x_0) + \frac{4}{3}\sum_{i=0}^{n-1}\underline{f}(x_{2i+1}) + \frac{2}{3}\sum_{i=1}^{n}\underline{f}(x_{2i})\frac{1}{3}\underline{f}(x_n)\right] + E\left(\underline{f}\right), \tag{8.14}$$

$$I\left(\overline{f}\right) = h\left[\frac{1}{3}\overline{f}(x_0) + \frac{4}{3}\sum_{i=0}^{n-1}\overline{f}(x_{2i+1}) + \frac{2}{3}\sum_{i=1}^{n}\overline{f}(x_{2i})\frac{1}{3}\overline{f}(x_n)\right] + E\left(\overline{f}\right),$$

where

$$E\left(\underline{f}\right) = -\frac{h^2}{180}(b-a)\underline{f}^{(4)}\left(\underline{\zeta}\right), \tag{8.15}$$

$$E\left(\overline{f}\right) = -\frac{h^4}{180} = (b-a)\overline{f}^{(4)}\left(\overline{\zeta}\right), \quad \underline{\zeta}, \overline{\zeta} \in [x_0, x_n]$$

8.6 Example

Consider the following interval integral:

$$\int_0^1 \tilde{k}x^2 \, dx, \quad \tilde{k} = [-1, 1].$$

the exact solution is $\frac{1}{3}[-1, 1]$.

From Trapezoidal rule with $h = 1$:

$$Q\left(\underline{f}\right) = \frac{1}{2}(-1), \quad Q\left(\overline{f}\right) = \frac{1}{2}(1),$$

$$\underline{f}'' = 2(-1), \quad \overline{f}'' = 2(1)$$

and

$$E\left(\underline{f}\right) = -\frac{1}{6}(1), \quad E\left(\overline{f}\right) = -\frac{1}{6}(-1),$$

now with $h = \frac{1}{2}$:

$$I\left(\underline{f}\right) = \frac{3}{8}(-1), \quad I\left(\overline{f}\right) = \frac{3}{8}(1),$$

$$E\left(\underline{f}\right) = \frac{1}{24}(-1), \quad E\left(\overline{f}\right) = \frac{1}{24}(1)$$

8.7 Example

Consider the following interval integral:

$$\int_{-1}^{-2} \tilde{k}x\,dx, \quad \tilde{k}=[0,2],$$

the exact solution is $\frac{3}{2}[-2,0]$.

From trapezoidal rule with $h=1$:

$$Q\left(\underline{f}\right)=\frac{3}{2}(-2), \quad Q\left(\overline{f}\right)=\frac{3}{2}(0), E\left(\underline{f}\right)=E\left(\overline{f}\right)=(0)$$

9

Gauss Integration

9.1 Gaussian Integration

This method is another method for approximating the area under the curve. The general form of this rule is the same as Newton-Cotes rule. One of the advantages of this method over Newton-Cotes method is the application of orthogonal functions, so in the corresponding linear system of equations of this method, the matrix of coefficients is an orthogonal matrix, that is, all rows and columns are perpendicular to each other. So, it has all the advantages of orthogonal systems. Also, in some of these methods, the absolute magnitude of the weights is less than or equal to one, in which case, we will not have the product error propagation. Another advantage of this method is that we do not need to solve the system to obtain integration weights, because they can be obtained through one formula. Regarding the points of integration, it can be said that they are the roots of the orthogonal polynomials that are used in the integration method, that is, if we use the Legendre orthogonal polynomials, the Gauss Legendre rule is obtained, and if we use Hermit orthogonal polynomials, Gaussian Hermit rule is obtained, and if we use Laguerre orthogonal polynomials, we will have Gauss-Laguerre method. Due to the fact that in this method, in addition to integration weights, integration points are also unknown, so the number of unknowns in this method will be twice that of Newton-Cotes methods. So, the basis of the used space will have more members. Therefore, the integration method will be more accurate for higher-order polynomials. That is, the accuracy of integration increases. Gauss quadrature rule can be summarized as follows:

$$\int_a^b w(x)f(x)dx = \sum_{j=1}^n H_j f(a_j) + E$$

$$H_j = -\frac{a_{n+1}\gamma_n}{a_n Q_{n+1}(a_j)Q_n'(a_j)}$$

DOI: 10.1201/9781003218173-9

$$E = \frac{\gamma_n}{a_n^2 (2n)!} f^{(2n)}(\eta), \quad \eta \in (a,b)$$

$$Q_n(a_j) = 0, \quad j = 1,\ldots,n$$

In the above relations:

$$\gamma_n = \int_a^b w(x) Q_n^2(x) dx$$

We examine the different methods in the following.

9.1.1 Gauss Legendre

In this method, the integration interval is $[-1,1]$ and $w(x) = 1$.
Then

$$P_{n+1}(x) = \frac{2n+1}{n+1} x P_n(x) - \frac{n}{n+1} P_{n-1}(x), \quad P_0(x) = 1$$

$$\gamma_n = \frac{2}{2n+1}, \quad a_n = \frac{(2n)!}{2^n (n!)^2}$$

$$H_j = -\frac{2}{(n+1) P_{n+1}(a_j) P_n'(a_j)}$$

$$E = \frac{2^{2n+1}(n!)^4}{(2n+1)((2n)!)^3} f^{(2n)}(\eta), \quad \eta \in (-1,1)$$

9.1.2 Problem

Obtain Gauss-Legendre integration weights and calculate the method error value.

Solution: For Legendre orthogonal polynomials, we have:

$$P_{n+1}(x) = \frac{2n+1}{n+1} x P_n(x) - \frac{n}{n+1} P_{n-1}(x)$$

$$P_n(x) = \frac{1}{2^n \cdot n!} \cdot \frac{d^n}{dx^n} (x^2 - 1)^n$$

$$\int_{-1}^1 P_m(x) \cdot P_n(x) dx = \begin{cases} 0 & m \neq n \\ \dfrac{2}{2n+1} & m = n \end{cases}$$

we have:

$$H_k = -\frac{\dfrac{2(2n+2)!}{2^{2n+1}(2n+1)((n+1)!)^2}}{\dfrac{(2n)!}{2^n \cdot (n!)^2} \cdot P'_{n+1}(x_k) P_n(x_k)} = -\frac{2}{(n+1)P'_{n+1}(x_k) P_n(x_k)}$$

where x_k s are the roots of $P_n(x) = 0$.

To calculate the error, we proceed as following

$$P_n(x) = \frac{1}{2^n \cdot n!} \cdot \frac{d^n}{dx^n}\left(x^2 - 1\right)^n$$

$$= \frac{1}{2^n \cdot n!} \cdot \frac{d^n}{dx^n}\left(x^{2n} + \cdots\right)$$

$$= \frac{1}{2^n \cdot n!} 2n \cdot (2n-1)\ldots(2n-n)x^{2n-n} + \cdots$$

$$= \frac{(2n)!}{2^n \cdot (n!)} x^n + \cdots$$

where $\dfrac{(2n)!}{2^n \cdot (n!)^2}$ is the coefficient of the leading sentence in $P_n(x)$.

If $F(x) = \prod_{j=1}^{n}(x - x_j)$, then $P_n(x) = a_n F(x)$. According to the error relation, we have:

$$E = \frac{f^{(2n)}(\eta)}{(2n)!} \int_a^b w(x) F^2(x) dx$$

Therefore, for Gauss-Legendre error we will have:

$$E = \frac{f^{(2n)}(\eta)}{(2n)!} \int_{-1}^{1} \frac{2^{2n} \cdot (n!)^4}{((2n)!)^2} P_n^2(x) dx$$

Because $\displaystyle\int_{-1}^{1} P_n^2(x) dx = \frac{2}{2n+1}$, then

$$E = \frac{2^{2n+1}(n!)^4}{(2n+1)((2n)!)^3} f(2n)(\eta)$$

9.1.3 Problem

The integral $\langle f, g \rangle = \int_{-1}^{1} f(x)g(x)dx$ defines the inner product of the functions $f, g \in c[-1,1]$. Show that if f and g are polynomials of at most degree $n-1$ and for $i = 1,\ldots,n$, x_is are the roots of Legendre polynomials of degree n and also $\gamma_i = \int_{-1}^{1} L_i(x)dx$, where $L_i(x) = \prod_{k=1,k\neq i}^{n} \frac{x-x_k}{x_i-x_k}$ for $i = 1,\ldots,n$, then:

$$\langle f, g \rangle = \sum_{i=1}^{n} \gamma_i f(x_i)g(x_i)$$

Solution: It suffices to show that the Gauss-Legendre quadrature rule is accurate for the function $f(x)g(x)$, that is,

$$\int_{-1}^{1} f(x)g(x)dx = \sum_{i=1}^{n} \gamma_i f(x_i)g(x_i)$$

and this is obvious from the error equation. Because $f \cdot g \in \Pi_{2n-2}$ and the error for polynomials of at most degree $2n - 1$ is zero.

9.1.4 Problem

Consider the following Legendre polynomials.

$$P_j(x) := \frac{j!}{(2j)!} \cdot \frac{d^j}{dx^j}(x^2 - 1)^j, \quad j = 0,1,\ldots$$

Show that the leading coefficient of $P_j(x)$ is equal to one.
Solution: We must show:

$$\frac{d^j}{dx^j}P_j(x) = j!$$

So, we have:

$$\frac{d^j}{dx^j}P_j(x) = \frac{j!}{(2j)!}\left(\frac{d^j}{dx^j}\left(\frac{d^j}{dx^j}(x^2-1)^j\right)\right)$$

$$= \frac{j!}{(2j)!}\left(\frac{d^{2j}}{dx^{2j}}(x^2-1)^j\right)$$

On the other hand

$$\left(x^2-1\right)^j = \sum_{k=0}^{j} C_j^k (-1)^k \left(x^2\right)^{j-k}$$

$$= \sum_{k=0}^{j} C_j^k (-1)^k \left(x^2\right)^{2j-2k}$$

Obviously, if we take the $(2j)$-th derivative from the above $(j+1)$ sentences, only the first sentence will be non-zero and the rest of the sentences will be zero, so

$$\frac{d^j}{dx^j} P_j(x) = \frac{j!}{(2j)!} \cdot (2j)! = j!$$

9.1.5 Problem

Consider the Gauss integration with $[a,b] = [-1,1]$ and $w(x) = 1$.

A) Suppose that

$$P_j(x) = (-1)^{j+1} P_j(-x)$$

Show that in the recursive relation

$$P_0(x) \equiv 1$$

$$P_{j+1}(x) \equiv \left(x - \delta_{j+1}\right) P_j(x) - \gamma_{j+1}^2 P_{j-1}(x) \quad j \geq 0$$

where

$$\delta_{j+1} := \frac{\langle xP_j, P_j \rangle}{\left(P_j, P_j\right)} \quad j \geq 0$$

$$\gamma_{j+1}^2 := \begin{cases} 1 & j = 0 \\ \dfrac{\langle P_j, P_j \rangle}{\langle P_{j-1}, P_{j-1} \rangle} & j \geq 1 \end{cases}$$

for orthogonal polynomials

$$P_j(x) := \frac{j!}{(2j)!} \cdot \frac{d^j}{dx^j} \left(x^2-1\right)^j, \quad j = 0,1,\ldots$$

δ_j is equal to zero for $j > 0$.

B) Show

$$I = \int_{-1}^{1} \left(x^2 - 1\right)^j dx = \frac{(-1)^j 2^{2j+1}}{(2j+1)C_{2j}^j}$$

C) Calculate the value of $\langle P_j, P_j \rangle$ using integration by part in part (b) and show that for $j > 0$, for the recursive relation defined in part (a), we have:

$$\gamma_{j+1}^2 = \frac{j^2}{(2j+1)(2j-1)}$$

Solution:

A) It is enough to prove that $\langle xP_{j-1}, P_{j-1} \rangle = 0$

$$\int_{-1}^{1} xP_{j-1}^2(x)dx = \int_{1}^{-1} -xP_{j-1}^2(-x)(-dx)$$

$$= -\int_{1}^{-1} xP_{j-1}^2(-x)dx$$

$$= -\int_{1}^{-1} x\left((-1)^{2j} P_{j-1}^2(-x)\right)dx$$

$$= -\int_{1}^{-1} xP_{j-1}^2(x)dx$$

In this case, we will have

$$2\int_{-1}^{1} xP_{j-1}^2(x)dx = 0$$

Then,

$$\langle xP_{j-1}, P_{j-1} \rangle = 0$$

Thus,

$$\delta_j = 0$$

B)

$$I = \int_{-1}^{1} \left(x^2 - 1\right)^j dx = \int_{-1}^{1} (x+1)^j (x-1)^j dx$$

Using the rule of integration by part and assuming:

$$u = (x-1)^j \Rightarrow du = j(x-1)^{j-1} dx$$

$$dv = (x+1)^j dx \Rightarrow v = \frac{(x+1)^{j+1}}{j+1}$$

we will have:

$$I = \left[(x-1)^j \frac{(x+1)^{j+1}}{j+1} \right]_{-1}^{1} = \int_{-1}^{1} \frac{j}{j+1} (x-1)^{j-1} (x+1)^{j+1} dx$$

In the above relation, the first sentence is zero. Using the integration by parts, again and assuming the following

$$u = (x-1)^{j-1} \Rightarrow du(j-1)(x-1)^{j-2} dx$$

$$dv = (x+1)^{j+1} dx \Rightarrow v = \frac{(x+1)^{j+2}}{j+2}$$

We can write:

$$I = -\frac{j}{j+1} \left[\frac{1}{j+2} (x-1)^{j-1} (x+1)^{j+2} \right]_{-1}^{1} + \frac{j}{j+1} \int_{-1}^{1} \frac{j-1}{j+2} (x-1)^{j-2} (x+1)^{j+2} dx$$

So,

$$I = \frac{(-1)^2 j(j-1)}{(j+1)(j+2)} \int_{-1}^{1} (x-1)^{j-2} (x+1)^{j+2} dx$$

By continuing the integration by part method, we will finally have

$$I = \frac{(-1)^j j(j-1)...1}{(j+1)(j+2)...(2j)} \int_{-1}^{1} (x+1)^{2j} dx$$

$$= \frac{(-1)^j j! \, 2^{2j+1}}{(j+1)...(2j)(2j+1)}$$

$$= \frac{(-1)^j 2^{2j+1}}{(2j+1)C_{2j}^j}$$

C) It is necessary to explain that for $1 < j$, polynomials $\dfrac{d^l}{dx^l}(x^2 - 1)^j$ will be divisible by $x^2 - 1$ and for $1 > 2j$, $\dfrac{d^l}{dx^l}(x^2 - 1)^j = 0$. We have:

$$\langle P_j, P_j \rangle = \int_{-1}^{1} P_j^2(x)dx$$

$$= \left(\frac{j!}{(2j)!} \right)^2 \int_{-1}^{1} \frac{d^j}{dx^j}(x^2 - 1)^j \cdot \frac{d^j}{dx^j}(x^2 - 1)^j \, dx$$

Using the integration by parts and assuming

$$u = \frac{d^j}{dx^j}(x^2 - 1)^j \Rightarrow du = \frac{d^{j+1}}{dx^{j+1}}(x^2 - 1)^j \, dx$$

$$dv = \frac{d^j}{dx^j}(x^2 - 1)^j \, dx \Rightarrow v = \frac{d^{j-1}}{dx^{j-1}}(x^2 - 1)^j$$

and by putting $a = \left(\dfrac{j!}{(2j)!} \right)^2$, we will have

$$\langle P_j, P_j \rangle = a \left[\frac{d^j}{dx^j}(x^2 - 1)^j \cdot \frac{d^{j-1}}{dx^{j-1}}(x^2 - 1)^j \right]_{-1}^{1}$$

$$- a \int_{-1}^{1} \frac{d^{j-1}}{dx^{j-1}}(x^2 - 1)^j \cdot \frac{d^{j+1}}{dx^{j+1}}(x^2 - 1)^j \, dx \tag{9.1}$$

The first sentence of the equation (9.1) is zero. By reusing the integration by parts method, we can write:

$$\langle P_j, P_j \rangle = -a \left[\frac{d^{j+1}}{dx^{j+1}}(x^2 - 1)^j \cdot \frac{d^{j-2}}{dx^{j-2}}(x^2 - 1)^j \right]_{-1}^{1}$$

$$+ a \int_{-1}^{1} \frac{d^{j-2}}{dx^{j-2}}(x^2 - 1)^j \cdot \frac{d^{j+2}}{dx^{j+2}}(x^2 - 1)^j \, dx$$

So,

$$\langle P_j, P_j \rangle = a \int_{-1}^{1} \frac{d^{j-2}}{dx^{j-2}}(x^2 - 1)^j \cdot \frac{d^{j+2}}{dx^{j+2}}(x^2 - 1)^j \, dx$$

By continuing the integration by part method, we will finally have

$$\langle P_j, P_j \rangle = (-1)^j \cdot a \int_{-1}^{1} (x^2 - 1)^j \frac{d^{2j}}{dx^{2j}} (x^2 - 1)^j \, dx$$

$$= (-1)^j \cdot a \cdot (2j)! \frac{(-1)^j \cdot 2^{2j+1}}{(2j+1)C_{2j}^j}$$

$$= \frac{\left(\dfrac{j!}{(2j)!}\right)^2 \cdot 2^{2j+1} \cdot (2j)!}{\dfrac{(2j)!}{(j!)^2} \cdot (2j+1)}$$

$$= \frac{(j!)^4 \cdot 2^{2j+1}}{\left((2j)!\right)^2 \cdot (2j+1)}$$

Therefore, it can be concluded

$$\gamma_{j+1}^2 = \frac{\langle P_j, P_j \rangle}{\langle P_{j-1}, P_{j-1} \rangle}$$

$$= \frac{j^4 \cdot 2^2 \cdot (2j-1)}{(2j)^2 \cdot (2j-1)^2 \cdot (2j+1)}$$

$$= \frac{j^2}{(2j-1)(2j+1)}$$

9.1.6 Gauss Laguerre

In this method, the integration interval is $[0, \infty)$ and $w(x) = e^{-x}$. Then,

$$L_{n+1}(x) = (1 + 2n - x)L_n(x) - n^2 L_{n-1}(x), \quad L_0(x) = 1$$

$$\gamma_n = (n!)^2, \quad a_n = (-1)^n$$

$$H_j = \frac{(n!)^2}{L_{n+1}(a_j)L_n'(a_j)}$$

$$E = \frac{(n!)^2}{((2n)!)^2} f^{(2n)}(\eta), \quad \eta \in [0, \infty]$$

9.1.7 Gauss Hermite

In this method, the integration interval is $(-\infty,\infty)$ and $w(x)=e^{-x^2}$.
Then,

$$H_{n+1}(x)=2xH_n(x)-2nH_{n-1}(x), \quad H_0(x)=1$$

$$\gamma_n=2^n n!\sqrt{\pi}, \quad a_n=2^n$$

$$H_j=-\frac{2^{n+1}n!\sqrt{\pi}}{H_{n+1}(a_j)H_n'(a_j)}$$

$$E=\frac{n!\sqrt{\pi}}{2^n(2n)!}f^{(2n)}(\eta), \quad \eta\in(-\infty,\infty)$$

9.2 Gauss-Kronrod Quadrature Rule

In mathematics, the topic of approximation has many applications. The Gauss-Kronrod quadrature method is another method for approximating the integration based on the roots of Stieltjes and orthogonal polynomials. This method was introduced in 1964 by the Russian mathematician Kronrod. In this section, we first introduce Stieltjes polynomial and then explain the reason of the positiveness of the weights related to its nodes and finally, examine a special weight function. For this purpose, we first introduce and briefly review the required concepts and definitions.

As mentioned earlier, the orthogonal concept is naturally associated with quadrature and increases its degree of accuracy. The quadrature rules depend on the non-negative weight function $w(x)$ on the interval $[a,b]$, so that the torques $\mu_k=\langle x^k,1\rangle=\int_a^b w(x)x^k\,dx$ are present and bounded.

As we know, the Gaussian quadrature rule for integral $I(f)=\int_a^b w(x)f(x)dx$

is $Q_n(f)=\sum_{k=1}^n f(x_k)H_k$ with points $a<x_1<\cdots<x_n<b$, and we choose the weights associated with it (H_ks) in such a way that for each function of at most degree $2n-1$, such as f we have $Q_n(f)=I(f)$. We will now generalize this method. The rule of quadrature with the highest possible degree of accuracy is in the form of

$$\int_a^b w(x)f(x)dx \approx G(n,n+1) = \sum_{k=1}^n H_k f(\xi_k) + \sum_{j=1}^{n+1} B_j f(x_j) \qquad (9.2)$$

where the real function f is integrable on the interval $[a,b]$ and has the weight function $w(x)$ (with a positive sign) and contains n predetermined points and $n+1$ free points. The most interesting case occurs when the predetermined points are the points of the well-known Gaussian quadrature rule (such as the Chebyshev and Legendre quadrature rule, etc.) because otherwise it may lead to divergence.

$G(n,n+1)$ is a linear function on the interval $[a,b]$. The interesting thing about this method is that it mixes n roots with other $n+1$ roots, so that the rule error is estimated to the desired level, which means that in addition to the definition of $\xi_k \in [a,b]$, x_js must also be selected from the interval $[a,b]$. Note that by "point mixing" we mean that Gaussian points are between free points.

We obtain the weights H_k and B_j and all points x_j $(n+2(n+1) = 3n+2$ unknown parameters) in such a way that the quadrature rule is accurate for all polynomials of at most degree $3n+1$. For $j = 1,\ldots,n+1$, $x_j's$ are the roots of Stieltjes polynomial E_{n+1} of degree $n+1$. These polynomials satisfy the orthogonal condition:

$$\int_a^b w(x)p_n(x)E_{n+1}(x)x^k \, dx = 0$$

$$E(f) = -\frac{h^2}{180}(b-a)f^{(4)}(\zeta), \qquad (9.3)$$

$$E(f) = -\frac{h^2}{180}(b-a)f^{(4)}(\zeta), \quad \zeta \in [x_0, x_n]$$

9.3 Gaussian Quadrature for Approximate of Interval Integrals

Let f be an interval function. For any natural number n, the Gaussian formulas are

$$\int_a^b f(x)w(x)dx = \sum_{i=0}^n A_i f_i + E, \quad f_i = f(x_i), \quad i = 0,\ldots,n \qquad (9.4)$$

which provide approximate values for $\int_a^b f(x)w(x)dx$. The interval form of (9.4) is as follows:

$$\int_a^b \underline{f}(x)w(x)dx = \sum_{i=1}^n A_i \underline{f}(x_i) + E\left(\underline{f}\right), \tag{9.5}$$

$$\int_a^b \overline{f}(x)w(x)dx = \sum_{i=1}^n A_i \overline{f}(x_i) + E\left(\overline{f}\right),$$

The weights $A_i, i = 0,\ldots,n$, are real numbers with the property $\sum_{i=1}^n A_i = \int_a^b w(x)dx$. This follows (9.5) when applied to $\underline{f}(x) = \overline{f}(x) = 1$.

Let

$$Q\left(\underline{f}\right) = \sum_{i=1}^n A_i \underline{f}(x_i), \tag{9.6}$$

$$Q\left(\overline{f}\right) = \sum_{i=1}^n A_i \overline{f}(x_i),$$

Thus from (9.5), we have

$$I\left(\underline{f}\right) = \int_a^b \underline{f}(x)w(x)dx = Q\left(\underline{f}\right) + \left(\underline{f}\right), \tag{9.7}$$

$$I\left(\overline{f}\right) = \int_a^b \overline{f}(x)w(x)dx = Q\left(\overline{f}\right) + \left(\overline{f}\right),$$

Let $\underline{f} \in C^{2n}[a,b], \overline{f} \in C^{2n}[a,b]$ it can be shown that the approximation error may be expressed as follows:

$$E\left(\underline{f}\right) = \frac{\underline{f}^{2n}(\eta)}{(2n)!} \int_a^b w(x)q^2(x)dx = K\underline{f}^{2n}(\eta), \quad \underline{\eta} \in [a,b] \tag{9.8}$$

$$E\left(\overline{f}\right) = \frac{\overline{f}^{2n}(\overline{\eta})}{(2n)!} \int_a^b w(x)q^2(x)dx = K\overline{f}^{2n}(\overline{\eta}), \quad \overline{\eta} \in [a,b]$$

where $K = \dfrac{1}{(2n)!} \displaystyle\int_a^b w(x)q^2(x)dx$ and $q(x) = \displaystyle\prod_{i=0}^{n}(x - x_i)$ that x_i, $i = 0,\ldots,n$ are the toots of Legendre, Chebyshev, and Laguerre polynomials. The following theorem is proving that $Q(\underline{f}), Q(\overline{f})$ coverage to $I(\underline{f}), I(\overline{f})$, respectively.

Note

We assume that the arbitrary derivatives of the function f at different points in its domain are defined as intervals.

9.4 Gauss-Legendre Integration Rules for Interval Valued Functions

In these rules, we have $w(x) = 1$, $[a,b] = [-1,1]$. Thus, (9.4) is as follows:

$$\int_{-1}^{1} \underline{f}(x)dx = \sum_{i=0}^{n} A_i \underline{f}(x_i) + E(\underline{f}), \tag{9.9}$$

$$\int_{-1}^{1} \overline{f}(x)dx = \sum_{i=0}^{n} A_i \overline{f}(x_i) + E(\overline{f}),$$

where

$$E(\underline{f}) = \frac{\underline{f}^{2n}(\eta)}{(2n)!} \int_{-1}^{1} q^2(x)dx, \quad \underline{\eta} \in [-1,1], \tag{9.10}$$

$$E(\overline{f}) = \frac{\overline{f}^{2n}(\overline{\eta})}{(2n)!} \int_{-1}^{1} q^2(x)dx, \quad \overline{\eta} \in [-1,1],$$

where $q(x) = \displaystyle\prod_{i=0}^{n}(x - x_i)$ that x_i, $i = 0,\ldots,n$ are the roots of Legendre polynomials. In these methods, $\displaystyle\sum_{i=1}^{n} A_i = 2$.

9.4.1 One-Point Gauss-Legendre Integration Rule

$$I(\underline{f}) = 2\underline{f}(0) + E(\underline{f}), \tag{9.11}$$

$$I(\overline{f}) = 2\overline{f}(0) + E(\overline{f}),$$

where

$$E\left(\underline{f}\right) = \frac{1}{3}\underline{f}''\left(\underline{\eta}\right),$$

$$E\left(\overline{f}\right) = \frac{1}{3}\overline{f}''\left(\overline{\eta}\right), \quad \underline{\eta}, \overline{\eta} \in [-1,1]$$

9.4.2 Two-Point Gauss-Legendre Integration Rule

$$I\left(\underline{f}\right) = \underline{f}\left(-\frac{1}{\sqrt{3}}\right) + \underline{f}\left(\frac{1}{\sqrt{3}}\right) + E\left(\underline{f}\right), \qquad (9.12)$$

$$I\left(\overline{f}\right) = \overline{f}\left(-\frac{1}{\sqrt{3}}\right) + \overline{f}\left(\frac{1}{\sqrt{3}}\right) + E\left(\overline{f}\right).$$

where

$$E\left(\underline{f}\right) = \frac{1}{135}\underline{f}^{(4)}\left(\underline{\eta}\right),$$

$$E\left(\overline{f}\right) = \frac{1}{135}\overline{f}^{(4)}\left(\overline{\eta}\right), \quad \underline{\eta}, \overline{\eta} \in [-1,1]$$

9.4.3 Three-Point Gauss-Legendre Integration Rule

$$I\left(\underline{f}\right) = \frac{5}{9}\underline{f}\left(-\sqrt{\frac{3}{5}}\right) + \frac{8}{9}\underline{f}(0) + \frac{8}{9}\underline{f}\left(\sqrt{\frac{3}{5}}\right) + E\left(\underline{f}\right), \qquad (9.13)$$

$$I\left(\overline{f}\right) = \frac{5}{9}\overline{f}\left(-\sqrt{\frac{3}{5}}\right) + \frac{8}{9}\overline{f}(0) + \frac{8}{9}\overline{f}\left(\sqrt{\frac{3}{5}}\right) + E\left(\overline{f}\right),$$

where

$$E\left(\underline{f}\right) = \frac{1}{15750}\underline{f}^{(6)}\left(\underline{\eta}\right),$$

$$E\left(\overline{f}\right) = \frac{1}{15750}\overline{f}^{(6)}\left(\overline{\eta}\right), \quad \underline{\eta}, \overline{\eta} \in [-1,1]$$

9.5 Gauss-Chebyshev Integration Rules for Interval Valued Functions

In these rules, we have $w(x) = \dfrac{1}{\sqrt{1-x^2}}$, $[a,b] = [-1,1]$

we have

$$\int_{-1}^{1} \frac{\underline{f}(x)}{\sqrt{1-x^2}} dx = \sum_{i=1}^{n} A_i \underline{f}(x_i) + E(\underline{f}), \tag{9.14}$$

$$\int_{-1}^{1} \frac{\overline{f}(x)}{\sqrt{1-x^2}} dx = \sum_{i=1}^{n} A_i \overline{f}(x_i) + E(\overline{f}),$$

where

$$E(\underline{f}) = \frac{\underline{f}^{2n}(\underline{\eta})}{(2n)!} \int_{-1}^{1} \frac{q^2(x)}{\sqrt{1-x^2}} dx, \quad \underline{\eta} \in [-1,1], \tag{9.15}$$

$$E(\overline{f}) = \frac{\overline{f}^{2n}(\overline{\eta})}{(2n)!} \int_{-1}^{1} \frac{q^2(x)}{\sqrt{1-x^2}} dx, \quad \overline{\eta} \in [-1,1],$$

where $q(x) = \displaystyle\prod_{i=0}^{n} (x - x_i)$ that x_i, $i = 0,\dots,n$ are the roots of Chebyshev poly-

nomials. In these methods, $\displaystyle\sum_{i=1}^{n} A_i = \pi$.

9.5.1 One-Point Gauss-Chebyshev Integration Rule

$$I(\underline{f}) = \pi \underline{f}(0) + E(\underline{f}), \tag{9.16}$$

$$I(\overline{f}) = \pi \overline{f}(0) + E(\overline{f}),$$

where

$$E(\underline{f}) = \frac{1}{2} \underline{f}''(\underline{\eta}) \int_{-1}^{1} \frac{x^2}{\sqrt{1-x^2}} dx, \quad \underline{\eta} \in [-1,1],$$

$$E(\overline{f}) = \frac{1}{2} \overline{f}''(\overline{\eta}) \int_{-1}^{1} \frac{x^2}{\sqrt{1-x^2}} dx, \quad \overline{\eta} \in [-1,1],$$

9.5.2 Two-Point Gauss-Chebyshev Integration Rule

$$I\left(\underline{f}\right) = \frac{\pi}{2}\underline{f}\left(-\frac{\sqrt{2}}{2}\right) + \frac{\pi}{2}\underline{f}\left(\frac{\sqrt{2}}{2}\right) + E\left(\underline{f}\right) \tag{9.17}$$

$$I\left(\overline{f}\right) = \frac{\pi}{2}\overline{f}\left(-\frac{\sqrt{2}}{2}\right) + \frac{\pi}{2}\overline{f}\left(\frac{\sqrt{2}}{2}\right) + E\left(\overline{f}\right)$$

where

$$E\left(\underline{f}\right) = \frac{\underline{f}^{(4)}\left(\eta\right)}{4!} \int_{-1}^{1} \frac{\left(x^2 - \frac{1}{2}\right)^2}{\sqrt{1-x^2}}\,dx, \quad \underline{\eta} \in [-1,1],$$

$$E\left(\overline{f}\right) = \frac{\overline{f}^{(4)}\left(\eta\right)}{4!} \int_{-1}^{1} \frac{\left(x^2 - \frac{1}{2}\right)^2}{\sqrt{1-x^2}}\,dx, \quad \overline{\eta} \in [-1,1].$$

9.6 Gauss-Laguerre Integration Rules for Interval Valued Functions

In these rules, we have $w(x) = e^{-x}, [a,b] = [0,\infty)$. Thus,

$$\int_0^\infty \underline{f}(x)e^{-x}\,dx = \sum_{i=0}^{n} A_i \underline{f}(x_i) + E\left(\underline{f}\right)$$

$$\int_0^\infty \overline{f}(x)e^{-x}\,dx = \sum_{i=0}^{n} A_i \overline{f}(x_i) + E\left(\overline{f}\right)$$

where

$$E\left(\underline{f}\right) = \frac{\underline{f}^{(2n)}\left(\eta\right)}{(2n)!} \int_0^\infty q^2(x)e^{-x}\,dx, \quad \underline{\eta} \in [-1,1]$$

$$E\left(\overline{f}\right) = \frac{\overline{f}^{(2n)}\left(\overline{\eta}\right)}{(2n)!} \int_0^\infty q^2(x)e^{-x}\,dx, \quad \overline{\eta} \in [-1,1]$$

where $q(x) = \prod_{i=0}^{n}(x - x_i)$ that x_i, $i = 0,\ldots,n$ are the roots of Laguerre polyno-

mials. In these methods, $\sum_{i=0}^{n} A_i = 1$.

9.6.1 One-Point Gauss-Laguerre Integration Rule

$$I\left(\underline{f}\right) = \underline{f}(1) + E\left(\underline{f}\right),$$

$$I\left(\overline{f}\right) = \overline{f}(1) + E\left(\overline{f}\right).$$

where

$$E\left(\underline{f}\right) = \frac{1}{2}\underline{f}''(\eta)\int_0^\infty (x - 1)^2 e^{-x}\, dx,$$

$$E\left(\overline{f}\right) = \frac{1}{2}\overline{f}''(\overline{\eta})\int_0^\infty (x - 1)^2 e^{-x}\, dx, \quad \underline{\eta}, \overline{\eta} \in [0,\infty).$$

9.6.2 Two-Point Gauss-Laguerre Integration Rule

$$I\left(\underline{f}\right) = 0.853553390593 \cdot \underline{f}(x_0) + 0.146446609407\ \underline{f}(x_1) + E\left(\underline{f}\right)$$

$$I\left(\overline{f}\right) = 0.853553390593\ \overline{f}(x_0) + 0.146446609407\ \overline{f}(x_1) + E\left(\overline{f}\right).$$

$$x_0 = 0.585786437627,\ x_1 = 3.414213562373 \tag{9.18}$$

where

$$E\left(\underline{f}\right) = \frac{\underline{f}^{(4)}(\eta)}{(4)!}\int_0^\infty (x - x_0)^2 (x - x_1)^2 e^{-x}\, dx,$$

$$\underline{\eta}, \overline{\eta} \in [0,\infty)$$

$$E\left(\overline{f}\right) = \frac{\overline{f}^{(4)}(\overline{\eta})}{(4)!}\int_0^\infty (x - x_0)^2 (x - x_1)^2 e^{-x}\, dx,$$

9.7 Gaussian Multiple Integrals Method

Let f be an interval function. For any natural number n, the Gaussian formulas are as follows:

$$\int_c^d \int_a^b f(x,y)w(x,y)dxdy = \sum_{j=1}^m \sum_{i=1}^n A_j B_i f_{ij} + E \tag{9.19}$$

$$f_{ij} = f(x_i, y_j), \quad i = 1,\ldots,n, \ j = 1,\ldots,m.$$

Let $w(x,y) = w(x) \times w(y)$, which provide that approximate values for $\int_c^d \int_a^b f(x,y)w(x,y)dxdy$, the interval form of (9.19) is as follows:

$$\int_c^d \int_a^b \underline{f}(x,y)w(x,y)dxdy = \sum_{j=1}^m \sum_{i=1}^n A_j B_i \underline{f}(x_i, y_j) + E(\underline{f}), \tag{9.20}$$

$$\int_c^d \int_a^b \overline{f}(x,y)w(x,y)dxdy = \sum_{j=1}^m \sum_{i=1}^n A_j B_i \overline{f}(x_i, y_j) + E(\overline{f}),$$

The weights $A_j, B_i, j = 1,\ldots,m, i = 1,\ldots,n$, are positive real numbers with the property $\sum_{j=1}^m \sum_{i=1}^n A_j B_i = \int_c^d \int_a^b w(x,y)dxdy$.

This follows (9.4) when applied to $\underline{f}(x,y) = \overline{f}(x,y) = 1$. Let

$$Q(\underline{f}) = \sum_{j=1}^m \sum_{i=1}^n A_j B_i \underline{f}(x_i, y_j),$$

$$Q(\overline{f}) = \sum_{j=1}^m \sum_{i=1}^n A_j B_i \overline{f}(x_i, y_j),$$

we have

$$I(\underline{f}) = \int_c^d \int_a^b \underline{f}(x,y)w(x,y)dxdy = Q(\underline{f}) + E(\underline{f}),$$

$$I(\overline{f}) = \int_c^d \int_a^b \overline{f}(x,y)w(x,y)dxdy = Q(\overline{f}) + E(\overline{f})$$

Let $M = \{(x,y)|a \le x \le b, c \le y, \le d\}$. We assume that all needed partial derivatives exist and continuous on the M. It can be shown that the approximation error may be expressed as follows:

$$E(\underline{f}) = K_2 \sum_{i=1}^{n} B_i \frac{\partial^{2m} \underline{f}(x_i,\eta)}{\partial y^{2m}} + K_1 \sum_{j=1}^{m} C_j \frac{\partial^{2n} \underline{f}(\underline{\zeta},y_j)}{\partial x^{2n}} + K_1 K_2 \frac{\partial^{2(m+n)} \underline{f}(\underline{\zeta},\eta)}{\partial y^{2m} \partial x^{2n}},$$

$$\underline{\zeta} \in [a,b], \underline{\eta} \in [c,d]$$

$$E(\overline{f}) = K_2 \sum_{i=1}^{n} B_i \frac{\partial^{2m} \overline{f}(x_i,\overline{\eta})}{\partial y^{2m}} + K_1 \sum_{j=1}^{m} C_j \frac{\partial^{2n} \overline{f}(\overline{\zeta},y_j)}{\partial x^{2n}} + K_1 K_2 \frac{\partial^{2(m+n)} \overline{f}(\overline{\zeta},\overline{\eta})}{\partial y^{2m} \partial x^{2n}},$$

$$\overline{\zeta} \in [a,b], \overline{\eta} \in [c,d]$$

where $K_1 = \frac{1}{(2n)!} \int_a^b w(x) q^2(x) dx$, $K_2 = \frac{1}{(2n)!} \int_c^d w(y) q^2(y) dy$, $q(x) = \prod_{i=1}^{n} (x - x_i)$

and $q(y) = \prod_{j=1}^{m} (y - y_j)$ that $x_i, i = 1,\dots,n, y_j = 1,\dots,m$ are the roots of Legendre and Chebyshev polynomials. Let $K = [a,b] \times [c,d]$. The following theorem is proving that $Q(\underline{f}), Q(\overline{f})$ coverage to $I(\underline{f}), I(\overline{f})$, respectively.

9.8 Gauss-Legendre Multiple Integrals Rules for Interval Valued Functions

In these rules, we have $w(x)w(y) = 1, [a,b] = [c,d] = [-1,1]$.
Thus,

$$\int_{-1}^{1} \int_{-1}^{1} \underline{f}(x,y) dx dy = \sum_{j=1}^{m} \sum_{i=1}^{n} A_j B_i \underline{f}(x_i,y_j) + E(\underline{f}),$$

$$\int_{-1}^{1} \int_{-1}^{1} \overline{f}(x,y) dx dy = \sum_{j=1}^{m} \sum_{i=1}^{n} A_j B_i \overline{f}(x_i,y_j) + E(\overline{f}),$$

where

$$E\left(\underline{f}\right) = K_2 \sum_{i=1}^{n} B_i \frac{\partial^{2m} f\left(x_i, \underline{\eta}\right)}{\partial y^{2m}} + K_1 \sum_{i=1}^{m} C_j \frac{\partial^{2n} f\left(\underline{\zeta}, y_j\right)}{\partial x^{2n}} + K_1 K_2 \frac{\partial^{2(m+n)} f\left(\underline{\zeta}, \underline{\eta}\right)}{\partial y^{2m} \partial x^{2n}}$$

$$E\left(\overline{f}\right) = K_2 \sum_{i=1}^{n} B_i \frac{\partial^{2m} \overline{f}\left(x_i, \overline{\eta}\right)}{\partial y^{2m}} + K_1 \sum_{j=1}^{m} C_j \frac{\partial^{2n} \overline{f}\left(\overline{\zeta}, y_j\right)}{\partial x^{2n}} + K_1 K_2 \frac{\partial^{2(m+n)} \overline{f}\left(\overline{\zeta}, \overline{\eta}\right)}{\partial y^{2m} \partial x^{2n}}$$

$$\overline{\zeta}, \underline{\zeta}, \underline{\eta}, \overline{\eta} \in [-1, 1]$$

where $\quad K_1 = \dfrac{1}{(2n)!} \displaystyle\int_{-1}^{1} q^2(x) dx, \ K_2 = \dfrac{1}{(2m)!} \int_{-1}^{1} q^2(y) dy, \ q(x) = \displaystyle\prod_{i=1}^{n} (x - x_i)$ and

$q(y) = \displaystyle\prod_{j=1}^{m} (y - y_j)$ that $x_i, \ i = 1, \ldots, n, \ y_j = 1, \ldots, m$ are the roots of Legendre

polynomials. In these methods, $\displaystyle\sum_{j=1}^{m} \sum_{i=1}^{n} A_j B_i = 4$.

9.8.1 Composite One-Point Gauss-Legendre Integration Rule

$$I\left(\underline{f}\right) = 4\underline{f}(0,0) + E\left(\underline{f}\right),$$

$$I\left(\overline{f}\right) = 4\overline{f}(0,0) + E\left(\overline{f}\right),$$

where

$$E\left(\underline{f}\right) = \frac{2}{3} \frac{\partial^2 f(0, \underline{\eta})}{\partial y^2} + \frac{2}{3} \frac{\partial^2 f(\underline{\zeta}, 0)}{\partial x^2} + \frac{1}{9} \frac{\partial^4 f(\underline{\zeta}, \underline{\eta})}{\partial y^2 \partial x^2},$$

$$E\left(\overline{f}\right) = \frac{2}{3} \frac{\partial^2 \overline{f}(0, \overline{\eta})}{\partial y^2} + \frac{2}{3} \frac{\partial^2 \overline{f}(\overline{\zeta}, 0)}{\partial x^2} + \frac{1}{9} \frac{\partial^4 \overline{f}(\overline{\zeta}, \overline{\eta})}{\partial y^2 \partial x^2}$$

$$\overline{\zeta}, \underline{\zeta}, \underline{\eta}, \overline{\eta} \in [-1, 1]$$

9.8.2 Composite Two-Point Gauss-Legendre Integration Rule

$$I\left(\underline{f}\right) = \underline{f}\left(-\frac{1}{\sqrt{3}}, -\frac{1}{\sqrt{3}}\right) + \underline{f}\left(-\frac{1}{\sqrt{3}}, \frac{1}{\sqrt{3}}\right) + \underline{f}\left(\frac{1}{\sqrt{3}}, -\frac{1}{\sqrt{3}}\right) + \underline{f}\left(\frac{1}{\sqrt{3}}, \frac{1}{\sqrt{3}}\right) + E\left(\underline{f}\right)$$

$$I\left(\overline{f}\right) = \overline{f}\left(-\frac{1}{\sqrt{3}}, -\frac{1}{\sqrt{3}}\right) + \overline{f}\left(-\frac{1}{\sqrt{3}}, \frac{1}{\sqrt{3}}\right) + \overline{f}\left(\frac{1}{\sqrt{3}}, -\frac{1}{\sqrt{3}}\right) + \overline{f}\left(\frac{1}{\sqrt{3}}, \frac{1}{\sqrt{3}}\right) + E\left(\overline{f}\right)$$

where

$$
E\left(\underline{f}\right) = \frac{1}{135}\frac{\partial^2 \underline{f}^4\left(-\dfrac{1}{\sqrt{3}},\underline{\eta}\right)}{\partial y^4} + \frac{1}{135}\frac{\partial \underline{f}^4\left(\dfrac{1}{\sqrt{3}},\underline{\eta}\right)}{\partial y^4} + \frac{1}{135}\frac{\partial^1 \underline{f}^4\left(\underline{\zeta},-\dfrac{1}{\sqrt{3}}\right)}{\partial x^4}
$$

$$
+ \frac{1}{135}\frac{\partial^4 \underline{f}^1\left(\underline{\zeta},\dfrac{1}{\sqrt{3}}\right)}{\partial x^4} + \frac{1}{135}\frac{1}{135}\frac{\partial^8 \underline{f}\left(\underline{\zeta},\underline{\eta}\right)}{\partial y^4 \partial x^4}
$$

$$
E\left(\overline{f}\right) = \frac{1}{135}\frac{\partial \overline{f}^4\left(-\dfrac{1}{\sqrt{3}},\overline{\eta}\right)}{\partial y^4} + \frac{1}{135}\frac{\partial \overline{f}^4\left(\dfrac{1}{\sqrt{3}},\overline{\eta}\right)}{\partial y^4} + \frac{1}{135}\frac{\partial^1 \overline{f}^4\left(\overline{\zeta},-\dfrac{1}{\sqrt{3}}\right)}{\partial x^4}
$$

$$
+ \frac{1}{135}\frac{\partial^4 \overline{f}^1\left(\overline{\zeta},\dfrac{1}{\sqrt{3}}\right)}{\partial x^4} + \frac{1}{135}\frac{1}{135}\frac{\partial^8 \overline{f}\left(\overline{\zeta},\overline{\eta}\right)}{\partial y^4 \partial x^4}
$$

$$
\overline{\zeta},\underline{\zeta},\underline{\eta},\overline{\eta} \in [-1,1]
$$

9.8.3 Composite One- and Three-Point Gauss-Legendre Integration Rule

$$
I\left(\underline{f}\right) = \frac{10}{9}\underline{f}\left(0,-\sqrt{\frac{3}{5}}\right) + \frac{16}{9}\underline{f}(0,0) + \frac{10}{9}\underline{f}\left(0,\sqrt{\frac{3}{5}}\right) + E\left(\underline{f}\right)
$$

$$
I\left(\overline{f}\right) = \frac{10}{9}\overline{f}\left(0,-\sqrt{\frac{3}{5}}\right) + \frac{16}{9}\overline{f}(0,0) + \frac{10}{9}\overline{f}\left(0,\sqrt{\frac{3}{5}}\right) + E\left(\overline{f}\right)
$$

where

$$
E\left(\underline{f}\right) = \frac{2}{15750}\frac{\partial^6 \underline{f}^1\left(0,\underline{\eta}\right)}{\partial y^6} + \frac{5}{27}\frac{\partial^2 \underline{f}^1\left(\underline{\zeta},-\sqrt{\dfrac{3}{5}}\right)}{\partial y^2} + \frac{8}{27}\frac{\partial^2 \underline{f}\left(\underline{\zeta},0\right)}{\partial x^2}
$$

$$
+ \frac{5}{27}\frac{\partial^2 \underline{f}^1\left(\underline{\zeta},\sqrt{\dfrac{3}{5}}\right)}{\partial x^2} + \frac{1}{3}\frac{1}{15750}\frac{\partial^8 \underline{f}\left(\underline{\zeta},\underline{\eta}\right)}{\partial y^4 \partial x^4},
$$

$$E\left(\overline{f}\right) = \frac{2}{15750} \frac{\partial^6 \overline{f}^1(0,\overline{\eta})}{\partial y^6} + \frac{5}{27} \frac{\partial^2 \overline{f}^1\left(\overline{\zeta}, -\sqrt{\frac{3}{5}}\right)}{\partial y^2} + \frac{8}{27} \frac{\partial^2 \overline{f}\left(\overline{\zeta},0\right)}{\partial x^2}$$

$$+ \frac{5}{27} \frac{\partial^2 \overline{f}^1\left(\overline{\zeta},\sqrt{\frac{3}{5}}\right)}{\partial x^2} + \frac{1}{3} \frac{1}{15750} \frac{\partial^8 \overline{f}\left(\overline{\zeta},\overline{\eta}\right)}{\partial y^4 \partial x^4}$$

$$\overline{\zeta},\underline{\zeta},\underline{\eta},\overline{\eta} \in [-1,1]$$

9.9 Gauss-Chebyshev Multiple Integrals Rules for Interval Valued Functions

In these rules, we have $w(x) = \dfrac{1}{\sqrt{1-x^2}}$, $w(y) = \dfrac{1}{\sqrt{1-y^2}}$, $[a,b] = [c,d] = [-1,1]$.

Thus,

$$\int_{-1}^1 \int_{-1}^1 \frac{f(x,y)}{\sqrt{1-y^2}\sqrt{1-x^2}} \, dx\, dy = \sum_{j=1}^m \sum_{i=1}^n A_j B_i \underline{f}\left(x_i,y_j\right) + E\left(\underline{f}\right),$$

$$\int_{-1}^1 \int_{-1}^1 \frac{\overline{f}(x,y)}{\sqrt{1-y^2}\sqrt{1-x^2}} \, dx\, dy = \sum_{j=1}^m \sum_{i=1}^n A_j B_i \overline{f}\left(x_i,y_j\right) + E\left(\overline{f}\right),$$

where

$$E\left(\underline{f}\right) = K_2 \sum_{i=1}^n B_i \frac{\partial^{2m} f\left(x_i,\eta\right)}{\partial y^{2m}} + K_1 \sum_{i=1}^m C_i \frac{\partial^{2n} f\left(\zeta,y_j\right)}{\partial x^{2n}} + K_1 K_2 \frac{\partial^{2(m+n)} f\left(\zeta,\eta\right)}{\partial y^{2m}\partial x^{2n}}$$

$$E\left(\overline{f}\right) = K_2 \sum_{i=1}^n B_i \frac{\partial^{2m} \overline{f}\left(x_i,\overline{\eta}\right)}{\partial y^{2m}} + K_1 \sum_{j=1}^m C_j \frac{\partial^{2n} \overline{f}\left(\overline{\zeta},y_j\right)}{\partial x^{2n}} + K_1 K_2 \frac{\partial^{2(m+n)} \overline{f}\left(\overline{\zeta},\overline{\eta}\right)}{\partial y^{2m}\partial x^{2n}}$$

where $\quad K_1 = \dfrac{1}{(2n)!} \displaystyle\int_{-1}^1 \frac{q^2(x)}{\sqrt{1-x^2}} \, dx$, $K_2 = \dfrac{1}{(2m)!} \displaystyle\int_{-1}^1 \frac{q^2(y)}{\sqrt{1-y^2}} \, dy$, $q(x) = \displaystyle\prod_{i=1}^n (x-x_i)$

and $q(y) = \displaystyle\prod_{j=1}^m (y-y_j)$ that x_i, $i = 1,\ldots,n$, $y_j = 1,\ldots,m$ are the roots of Cheby-

shev polynomials. In this case, $\displaystyle\sum_{j=1}^m \sum_{i=1}^n A_j B_i = \pi^2$..

9.9.1 Composite One-Point Gauss-Chebyshev Integration Rule

$$I\left(\underline{f}\right) = \pi^2 \underline{f}(0,0) + E\left(\underline{f}\right),$$

$$I\left(\overline{f}\right) = \pi^2 \overline{f}(0,0) + E\left(\overline{f}\right),$$

where

$$E\left(\underline{f}\right) = \frac{\pi^2}{4} \frac{\partial^2 \underline{f}(0,\eta)}{\partial y^2} + \frac{\pi^2}{4} \frac{\partial^2 \underline{f}(\zeta,0)}{\partial x^2} + \frac{\pi^2}{16} \frac{\partial^4 \underline{f}(\zeta,\eta)}{\partial y^2 \partial x^2},$$

$$\zeta,\underline{\zeta},\eta,\overline{\eta} \in [-1,1]$$

$$E\left(\overline{f}\right) = \frac{\pi^2}{4} \frac{\partial^2 \overline{f}(0,\overline{\eta})}{\partial y^2} + \frac{\pi^2}{4} \frac{\partial^2 \overline{f}(\overline{\zeta},0)}{\partial x^2} + \frac{\pi^2}{16} \frac{\partial^4 \overline{f}(\overline{\zeta},\overline{\eta})}{\partial y^2 \partial x^2}$$

9.9.2 Composite One- and Two-Point Gauss-Chebyshev Integration Rule

$$I\left(\underline{f}\right) = \frac{\pi^2}{2} \underline{f}\left(0,-\frac{\sqrt{2}}{2}\right) + \frac{\pi^2}{2} \underline{f}\left(0,\frac{\sqrt{2}}{2}\right) + +E\left(\underline{f}\right)$$

$$E\left(\overline{f}\right) = \frac{\pi^2}{2} \overline{f}\left(0,-\frac{\sqrt{2}}{2}\right) + \frac{\pi^2}{2} \overline{f}\left(0,\frac{\sqrt{2}}{2}\right) + +E\left(\overline{f}\right)$$

where

$$E\left(\underline{f}\right) = \frac{\pi^2}{48} \frac{\partial^4 \underline{f}^1(0,\eta)}{\partial y^4} + \frac{\pi^2}{8} \frac{\partial^2 \underline{f}^1\left(\underline{\zeta},-\frac{\sqrt{2}}{2}\right)}{\partial x^2} + \frac{\pi^2}{8} \frac{\partial^2 \underline{f}^1\left(\underline{\zeta},\frac{\sqrt{2}}{2}\right)}{\partial x^2} + \frac{\pi^2}{192} \frac{\partial^6 \underline{f}(\zeta,\eta)}{\partial y^4 x^2},$$

$$E\left(\overline{f}\right) = \frac{\pi^2}{48} \frac{\partial^4 \overline{f}^1(0,\overline{\eta})}{\partial y^4} + \frac{\pi^2}{8} \frac{\partial^2 \overline{f}^1\left(\overline{\zeta},-\frac{\sqrt{2}}{2}\right)}{\partial x^2} + \frac{\pi^2}{8} \frac{\partial^2 \overline{f}^1\left(\overline{\zeta},\frac{\sqrt{2}}{2}\right)}{\partial x^2} + \frac{\pi^2}{192} \frac{\partial^6 \overline{f}(\overline{\zeta},\overline{\eta})}{\partial y^4 x^2}$$

$$\zeta,\underline{\zeta},\eta,\overline{\eta} \in [-1,1]$$

9.10 Composite Gauss-Legendre and Gauss-Chebyshev Integration Rule

In these rules, we have $w(x) = 1$, $w(y) = \dfrac{1}{\sqrt{1-y^2}}$, $[a,b] = [c,d] = [-1,1]$.

Thus,

$$\int_{-1}^{1}\int_{-1}^{1}\frac{f(x,y)}{\sqrt{1-y^2}}\,dx\,dy = \sum_{j=1}^{m}\sum_{i=1}^{n}A_jB_i\underline{f}(x_i,y_j)+E\left(\underline{f}\right),$$

$$\int_{-1}^{1}\int_{-1}^{1}\frac{\overline{f}(x,y)}{\sqrt{1-y^2}}\,dx\,dy = \sum_{j=1}^{m}\sum_{i=1}^{n}A_jB_i\overline{f}(x_i,y_j)+E\left(\overline{f}\right),$$

where

$$E\left(\underline{f}\right)= K_2\sum_{i=1}^{n}B_i\frac{\partial^{2m}\underline{f}(x_i,\underline{\eta})}{\partial y^{2m}} + K_1\sum_{i=1}^{m}C_j\frac{\partial^{2n}\underline{f}(\underline{\zeta},y_j)}{\partial x^{2n}} + K_1K_2\frac{\partial^{2(m+n)}\underline{f}(\underline{\zeta},\underline{\eta})}{\partial y^{2m}\partial x^{2n}}$$

$$E\left(\overline{f}\right)= K_2\sum_{i=1}^{n}B_i\frac{\partial^{2m}\overline{f}(x_i,\overline{\eta})}{\partial y^{2m}} + K_1\sum_{j=1}^{m}C_j\frac{\partial^{2n}\overline{f}(\overline{\zeta},y_j)}{\partial x^{2n}} + K_1K_2\frac{\partial^{2(m+n)}\overline{f}(\overline{\zeta},\overline{\eta})}{\partial y^{2m}\partial x^{2n}}$$

$$\overline{\zeta},\underline{\zeta},\underline{\eta},\overline{\eta}\in[-1,1]$$

where $q(x)=\displaystyle\prod_{i=1}^{n}(x-x_i)$ and $q(y)=\displaystyle\prod_{j=1}^{m}(y-y_j)$ that $x_i,\, i=1,\ldots,n, y_j=1,\ldots,m$

are the roots of Legendre and Chebyshev polynomials, respectively. In these

methods, $\displaystyle\sum_{j=1}^{m}\sum_{i=1}^{n}A_jB_i = 2\pi.$.

9.10.1 Composite One-Point Gauss-Legendre and One-Point Gauss-Chebyshev Multiple Integral Rule

$$I\left(\underline{f}\right)= 2\pi\underline{f}(0,0)+ E\left(\underline{f}\right),$$

$$I\left(\overline{f}\right)= 2\pi\overline{f}(0,0)+ E\left(\overline{f}\right),$$

where

$$E\left(\underline{f}\right)= \frac{\pi}{2}\frac{\partial^2 \underline{f}^1(0,\underline{\eta})}{\partial y^2} + \frac{\pi}{3}\frac{\partial^2 \underline{f}^1(\underline{\zeta},0)}{\partial x^2} + \frac{\pi}{12}\frac{\partial^4 \underline{f}(\underline{\zeta},\underline{\eta})}{\partial y^2\delta x^2},$$

$$E\left(\overline{f}\right)= \frac{\pi}{2}\frac{\partial^2 \overline{f}^1(0,\overline{\eta})}{\partial y^2} + \frac{\pi}{3}\frac{\partial^2 \overline{f}^1(\overline{\zeta},0)}{\partial x^2} + \frac{\pi}{12}\frac{\partial^4 \overline{f}(\overline{\zeta},\overline{\eta})}{\partial y^2\delta x^2}$$

$$\overline{\zeta},\underline{\zeta},\underline{\eta},\overline{\eta}\in[-1,1]$$

9.11 Adaptive Quadrature Rule

In this section, we first introduce the concept of Simpson's adaptive quadrature. Then, we obtain the Gauss-Lobatto three-point rule and the associated Kronrod expansion. Finally, we provide examples and, via examining the examples, we compare the efficiency of the Gauss-Lobatto-Kronrod adaptive integration method with that of other methods using the absolute error of these functions.

9.11.1 Introduction of Adaptive Quadrature Based on Simpson's Method

We know that the approximation of the integral by Simpson's method is as follows:

$$I = \int_a^b f(x)dx \approx S(a,b) = \frac{b-a}{6}\left(f(a) + 4f\left(\frac{a+b}{2}\right) + f(b)\right)$$

The error relation is as follows:

$$E\big(S(a,b)\big) = -\frac{(b-a)h^4}{180}f^{(4)}(\eta), \quad h = \frac{b-a}{2}, \quad \eta \in [a,b]$$

Now we want to develop the above approximation. For this purpose, we assume that $m = \frac{a+b}{2}$. In this case, we express I as follows

$$I = \int_a^m f(x)dx + \int_m^b f(x)dx$$

We have the following relation for approximation:

$$I \approx S(a,m) + S(m,b)$$

Now if we consider $f^{(4)}(\eta)$ with a constant number of approximately k, to estimate the error:

$$I \approx S(a,b) - \frac{(b-a)h^4}{180}k \tag{9.21}$$

and

$$I \approx S(a,m) - \frac{h\left(\frac{h}{2}\right)^4}{180}k + S(m,b) - \frac{h\left(\frac{h}{2}\right)^4}{180}k$$

$$= S(a,m) + S(a,m) - \frac{1}{16}\left(\frac{(b-a)h^4}{180}k\right)$$

or

$$16I = 16\big(S(a,m)+S(m,b)\big)-\frac{(b-a)h^4}{180}k \qquad\qquad (9.22)$$

we have:

$$15I = 16\big(S(a,b)+S(m,b)\big)-S(a,b)$$

Therefore, we conclude

$$I = \big(S(a,m)+S(m,b)\big) \approx \frac{1}{15}\big(S(a,m)+S(m,b)-S(a,b)\big)$$

We want to obtain an explicit estimate of the approximation error of the integral I by the Simpson's method for each half interval separately, and instead of calculating the difference between the actual and approximate values by the Simpson's method only once on the whole interval, the process is repeated if required. In general, it can be said that the method of adaptive quadrature is based on this fact.

Index